Intercultural Reciprocal Learning in Chinese and Western Education

Series Editors

Michael Connelly
Ontario Institute for Studies in Education
University of Toronto
Toronto, ON, Canada

Shijing Xu
Faculty of Education
University of Windsor
Windsor, ON, Canada

This book series grows out of the current global interest and turmoil over comparative education and its role in international competition. The specific series grows out of two ongoing educational programs which are integrated in the partnership, the University of Windsor-Southwest University Teacher Education Reciprocal Learning Program and the Shanghai-Toronto-Beijing Sister School Network. These programs provide a comprehensive educational approach ranging from preschool to teacher education programs. This framework provides a structure for a set of ongoing Canada-China research teams in school curriculum and teacher education areas. The overall aim of the Partnership program, and therefore of the proposed book series, is to draw on school and university educational programs to create a comprehensive cross-cultural knowledge base and understanding of school education, teacher education and the cultural contexts for education in China and the West.

George Zhou • Yuanrong Li • Jian Luo
Editors

Science Education and International Cross-Cultural Reciprocal Learning

Perspectives from the Nature Notes Program

palgrave
macmillan

Editors
George Zhou
Faculty of Education
University of Windsor
Windsor, ON, Canada

Yuanrong Li
College of Teacher Education
Southwest University
Chongqing, China

Jian Luo
Curriculum Innovation Center
Southwest University Affiliated
Secondary School
Chongqing, China

Intercultural Reciprocal Learning in Chinese and Western Education
ISBN 978-3-031-17156-7   ISBN 978-3-031-17157-4 (eBook)
https://doi.org/10.1007/978-3-031-17157-4

Cover illustration: beastfromeast/DigitalVision Vectors/Getty

This Palgrave Macmillan imprint is published by the registered company Springer Nature Switzerland AG.
The registered company address is: Gewerbestrasse 11, 6330 Cham, Switzerland

Series Foreword

The Series and East-West Contrasting Educational Narratives

This book series focuses on Chinese and Western education for the purpose of mutual understanding and reciprocal learning between the East and the West. The East has been a puzzle for the West, romanticized or demonized depending on the times. East-West relations have a long history of inquiry, and action has often been framed in competitive, ideological, and colonialist terms. In 1926 Dewey complained that "As far as we have gone at all, we have gone in loco parentis, with advice, with instruction, with example and precept. Like a good parent we would have brought up China in the way in which she should go" (p. 188). This "paternal" attitude, as Dewey called it, has not always been so benign. Economic, cultural and intellectual matters have often been in the forefront since the Opium Wars of the 19th Century. Intellectually the East-West dynamic is equally dramatic as found in works by authors such as Said (1978), Tu Wei-ming (1993), Hall and Ames (1999), Hayhoe and Pan (2001) and many others. These writers are part of a rich conceptual knowledge across cultures literature on the historical, philosophical, cultural and educational differences of the East and West.

Education is a vital topic of international discussion and essential component part of our global consciousness. Global discussions of economics, national and regional competition, and national and regional futures often turn to education. Meanwhile local educational discussions take place in social environments discourse of international awareness. 'How are our international neighbours doing?' 'How do they teach values?' 'We have to catch up.' These matters are vitally important. But they are not new. Higher education in universities and other forms of postsecondary education has occupied most of the attention. What is new, and what, in our view, is likely to have far-reaching impact, is the focus on school education and early childhood education as well as pre-service teacher education. For several reasons, not the least of which is national competition, the focus on school education has been driven by comparative achievement studies. When Shanghai school students topped the chart in the Program for International Student Assessment (PISA) studies the information was broadcast worldwide and generated ferocious discussion. One of the positive outcomes of this discussion is comparative research interest, the process of comparing educational similarities and differences in school practices, official policies, and social cultural influences. This comparative interest is all to the good and should help frame potential positive comparative futures. But comparative research on similarity and difference is not enough. We believe we need to reach beyond the study of similarities and differences and to explore life filled school practices of people in different cultures coming together and learning from one another. In this postmodern world of instant worldwide communication we need to go beyond comparative premises. Ideas, thoughts, images, research, knowledge, plans and policies are in constant interaction. This book series hopes to move our international educational research onto this collaborative and interactive educational landscape of schools, parents, communities, policy and international trends and forces.

Series Objectives and Contribution to Knowledge

The book series grew out of our seven-year Canada-China partnership study on reciprocal educational learning between Canada and China (Xu & Connelly, 2013–2020). The partnership developed from the current global interest and turmoil over comparative education and its role in international competition. The specific series grows out of two ongoing educational programs which are integrated in the partnership, the *University of Windsor-Southwest University Teacher Education Reciprocal Learning Program* and the *Shanghai-Toronto-Beijing Sister School Network*. These programs provide a comprehensive educational approach ranging from preschool to teacher education programs. This framework provides a structure for a set of ongoing Canada-China research teams in school curriculum and teacher education areas. The overall aim of the *Partnership* program, and therefore of the proposed book series, is to draw on school and university educational programs to create a comprehensive cross-cultural knowledge base and understanding of school education, teacher education and the cultural contexts for education in China and the West.

The first few books in the series will be direct outgrowths of our partnership study. But because of current global conditions, there is a great deal of important related work underway throughout the world. We encourage submissions to the series and expect the series to become a home for collaborative reciprocal learning educational work between the East and the West. The starting point in our Canada-China Reciprocal Learning Partnership's is the idea of a global community in which ideas, things, and people flow between countries and cultures (Xu & Connelly, 2013). There is intense public discussion in Canada over international relations with China. The publication of international student achievement scores that rank China at the top has resulted in growing scholarly and public discussion on the differences in our educational systems. The discussion tends to focus on economic and trade relations while educational reciprocity and reciprocal learning are often absent from educational discourse. Given that the Chinese are Canada's and Ontario's largest immigrant group and that Chinese students have statistically shown academic excellence, it is critical to explore what we can learn from Chinese philosophies of education and its educational system, and what Canada can offer China in return.

The Partnership's overall goal is to compare and contrast Canadian and Chinese education in such a way that the cultural narratives of each provide frameworks for understanding and appreciating educational similarities and differences. We expect other work generated outside our partnership Grant to have different starting points and socially relevant arguments. But we do expect all series works to share the *twin goals of mutual understanding and reciprocal learning.*

Built on these twin goals the purpose of the book series is to create and assemble the definitive collection of educational writings on the similarities, differences and reciprocal learnings between education in the East and the West. Drawing on the work of partnership oriented researchers throughout the world, the series is designed to:

- Build educational knowledge and understanding from a cross-cultural perspective;
- Support new approaches to research on curriculum, teaching and learning in schools and teacher education programs in response to change brought on by heightened global awareness;

- Provide a compelling theoretical frame for conceptualizing the philosophical and narrative historical trajectories of these two compelling worldviews on education, society and culture;
- Provide state of the art reviews of the comparative Chinese and English language literature on school curriculum and teacher education;
- Model, sustainable, school to school structures and methods of communication and educational sharing between Canada, other English speaking countries and China;
- Model, sustainable, structures and methods of initial teacher training in cross-cultural understanding;
- Contribute to a documented knowledge base of similarities, differences, comparisons and reciprocal learnings in elementary and secondary school teaching and learning curricula.

Michael Connelly
Professor Emeritus
Ontario Institute for Studies in Education
University of Toronto
Toronto, ON, Canada

Shijing Xu
Professor
Canada Research Chair
Faculty of Education
University of Windsor
Windsor, Canada

References

Dewey, J. (1926). *America and the Far East.* Survey, 1 May, 1926, 188. Later published in: John Dewey, The Later works, 1925-1953, (1984). Carbondale: Southern Illinois University Press. Volume 2: 1925–1927, pp. 1173–1175.

Hayhoe, R., & Pan, J. (Eds.). (2001). *Knowledge across cultures: A contribution to dialogue among civilizations.* Comparative Education Research Centre, The University of Hong Kong.

Said, E. W. (1978). *Orientalism.* Vintage Books.

Tu, W. (1993). *Way, learning, and politics: Essays on the Confucian intellectual.* State University of New York Press.

Xu, S., & Connelly F.M. (Project Directors) (2013). Reciprocal learning in teacher education and school education between Canada and China. Social Sciences and Humanities Research Council of Canada (SSHRC) 2013–2020 [Grant 895–2012-1011].

Praise for *Science Education and International Cross-Cultural Reciprocal Learning*

"This book describes an extraordinary reciprocal learning program that fostered significant intercultural learning between and among students and teachers in China and Canada. It affirms the importance of school-based curricula as a way to take Canadian students out of their dominated ethnocentric viewpoint and to more widely inform Chinese students about interdisciplinary ways of understanding science and technology. It tells a story about the challenges, successes, and excitement of doing international collaboration that should lead readers to develop their own reciprocal learning programs."

—Clayton Smith, *Professor, University of Windsor, Canada*

"This book is a product of a cross-culture reciprocal learning program between Canada and China. In this book, the readers will learn how students and teachers reflected on the relationship between human beings and nature and gained life aspiration from nature while they worked on their inquiry-based science projects through close observations of nature. Through this unique program, the sister school students and teachers learned different natural phenomena and customs of two countries and appreciated the biodiversity and cultural differences. Such an exchange program is a powerful platform for reciprocal learning!"

—Changchun Lin, *Professor, Chongqing Normal University, China*

Contents

Notes on Contributors

Xin Chen received a Master's degree in education from Southwest University. He was a graduate research assistant for the Canada-China nature notes reciprocal learning program. He currently is a science teacher in a public school in China.

Xue Chen received a Master's degree in education from Southwest University. She was a graduate research assistant for the Canada-China nature notes reciprocal learning program. She currently is a science teacher in a public school in China.

Shijie Deng received a Master's degree in education from Southwest University. She was a graduate research assistant for the Canada-China nature notes reciprocal learning program. She currently is a science teacher in a public school in China.

Han Deng received a Master's degree in education from Southwest University. She was a graduate research assistant for the Canada–China nature notes reciprocal learning program. She is currently a science teacher in a public school in China.

Liujun Guo received a Master's degree in education from Southwest University. She was a graduate research assistant for the Canada-China nature notes reciprocal learning program. She currently is a science teacher in a public school in China.

Shijing Guo received a Master's degree in education from Southwest University. She was a graduate research assistant for the Canada-China nature notes reciprocal learning program. She currently is a science teacher in a public school in China.

Sonia Ho received a Master's degree in education from the University of Windsor. She was a graduate research assistant for the Canada–China nature notes reciprocal learning program. She currently works as an educational assistant in public schools in Canada.

Banglong Jiang is a science teacher in Southwest University Affiliated Secondary school, China.

Yuanrong Li is a professor in the College of Teacher Education, Southwest University. China. She is the co-leader of the Canada–China nature notes reciprocal learning program. Her research areas include chemistry education, sustainability education, and teacher education. Her recent research work is focused on the reform of the Chemistry curriculum reform.

Yanan Liao received a Master's degree in education from Southwest University. She was a graduate research assistant for the Canada-China nature notes reciprocal learning program. She currently is a science teacher in a public school in China.

Yanhong Mao received a Master's degree in education from Southwest University. She was a graduate research assistant for the Canada-China nature notes reciprocal learning program. She currently is a science teacher in a public school in China.

Jin Mu received a Master's degree in education from Southwest University. She was a graduate research assistant for the Canada-China nature notes reciprocal learning program. She currently is a science teacher in a public school in China.

Yangxue Ou received a Master's degree in education from Southwest University. She was a graduate research assistant for the Canada–China nature notes reciprocal learning program. She is currently a science teacher in a public school in China.

Yiling Qi received a Master's degree in education from Southwest University. She was a graduate research assistant for the Canada-China nature notes reciprocal learning program. She currently is a science teacher in a public school in China.

Fei Tao received a Master's degree in education from Southwest University. He was a graduate research assistant for the Canada–China nature notes reciprocal learning program. He is currently a science teacher in a public school in China.

Yan Tian received a Master's degree in education from Southwest University. She was a graduate research assistant for the Canada–China nature notes reciprocal learning program. She is currently a science teacher in a public school in China.

Ruihan Wang received a Master's degree in education from Southwest University. She was a graduate research assistant for the Canada–China nature notes reciprocal learning program. She is currently a science teacher in a public school in China.

Han Yang received a Master's degree in education from Southwest University. She was a graduate research assistant for the Canada-China nature notes reciprocal learning program. She currently is a science teacher in a public school in China.

Qian Yang received a Master's degree in education from Southwest University. She was a graduate research assistant for the Canada–China nature notes reciprocal learning program. She is currently a science teacher in a public school in China.

Jiang Zeng received a Master's degree in education from Southwest University. She was a graduate research assistant for the Canada–China nature notes reciprocal learning program. She is currently a science teacher in a public school in China.

Guihua Zhang received a Master's degree in education from Southwest University. She was a graduate research assistant for the Canada-China nature notes reciprocal learning program. She currently is a science teacher in a public school in China.

George Zhou is a professor in the Faculty of Education, University of Windsor, Canada. He is the co-leader of the Canada–China nature notes reciprocal learning program. His research areas cover science education, environmental education, teacher education, parental involvement, comparative and international education. Cross-cultural learning and culture-responsive pedagogy are the main focuses of his recent research work.

List of Figures

List of Tables

CHAPTER 1

Introduction

George Zhou and Yuanrong Li

This book stems from the nature notes reciprocal learning program that took place in two sister schools between Canada and China. The program is designed and monitored by the science education research team within Xu and Connelly's (2013–2020) Canada-China Reciprocal Learning Partnership Program in School Education and Teacher Education funded by the Social Sciences and Humanities Research Council (SSHRC). The nature notes reciprocal learning program was collaboratively conducted between the University of Windsor, Canada, Southwest University (SWU), China, in partnership with the Prince Edward Elementary School, Windsor Ontario and the SWU Affiliated Secondary School, Chongqing, China.

Note: Some of the content included in this book was published first through a journal paper (Zhou et al., 2019).

G. Zhou (✉)
University of Windsor, Windsor, ON, Canada
e-mail: gzhou@uwindsor.ca

Y. Li
Southwest University, Chongqing, China
e-mail: hxliyr@swu.edu.cn

G. Zhou et al. (eds.), *Science Education and International Cross-Cultural Reciprocal Learning*, Intercultural Reciprocal Learning in Chinese and Western Education,
https://doi.org/10.1007/978-3-031-17157-4_1

The teachers from the two sister schools integrated the program into their curriculum. The students conducted nature observations in their local communities, and based on these observations, the students completed their projects step by step in a group or individually. Video conferencing through Skype regularly takes place between the schools for participating teachers and students to exchange ideas and share works. The nature notes reciprocal learning program does not aim at comparing students' achievements, but explore participating teachers' and students' international collaborative experience and the impact the program has on both teachers and students who participated in this unique teaching and learning journey. To this end, our research intends to respond to the two research questions:

1. What impacts does the nature notes reciprocal learning program bring to teachers and students in the Chinese-Canadian sister schools, and what did they learn from each other?
2. Were there any problems that both sides encountered and how did they handle those issues?

The Nature Notes Reciprocal Learning Program

The nature notes reciprocal learning program was designed as a school-based and science-related collaborative program, and it also carried a research plan that compared participating teachers' and students' respective experiences in the cross-cultural learning process and explored the impacts that they received. In such cases, it is necessary to understand what comparative education is about. Traditionally, comparative and international research mainly focuses on investigating the differences between the education systems of two or more countries and advocating how and what one country learns from another. It is more or less a static approach that mainly focuses on one-way knowledge transfer (Cowen, 2006). With the fast pace of globalization, the contact and collaboration between countries have increased dramatically in all areas of society, including education. This creates a new opportunity for comparative and international education move from statics approaches to dynamic approaches which address the mutual learning experiences in the context of interactions (Lewis et al., 2016). Researchers can explore differences, contradictions, negotiations, real-life changes, and mutual learning and benefits in the process of interactions between participating education organizations, educators, and learners (Carney, 2009). Reciprocity of learning is the key idea built into such a new perception of comparative and international education. Under this new perspective, cross-cultural learning needs to go beyond simple comparison in practices, values, and results and move to a paradigm that emphasizes a two-way learning process in the context of acting together (Connelly & Xu, 2019). As emphasized by Boud (1999), the two-way, reciprocal learning experiences should involve: "Participants learning from and with each other in both formal and informal ways. It includes mutual benefits and a sharing of knowledge, ideas and experience among participants" (p. 6).

Reciprocity of learning is the key idea built into a new perception of comparative and international education. The nature notes reciprocal learning program was designed under a framework of reciprocal learning. Through various deliberative work between the international research teams and the partner schools, nature notes were picked as a topic, activity, and framework for the partnership program between two sister schools. This science-based program was designed with a focus on experiential learning, without aiming to compare students' academic achievements. Given its nature of cross-national design, this program aimed to develop a partnership through which partner schools can learn from each other in practice, and to investigate the cross-cultural learning experiences of the participating students and teachers. Therefore, this program has a dual purpose: development and research.

As an educational program, the nature notes reciprocal learning program involved a combination of inside and outside classroom activities and blends the informal (outdoor observations) and formal learning experiences (in-class research) (Hofstein & Rosenfeld, 1996). Teachers from both sister schools integrated nature notes projects into their curricula. Following the selected solar terms, students from both sister schools conducted outdoor observations and collected data. Based on what they collected, students decided a topic to work on individually or with group members. Students also conducted in-class topic-related research and then produced a work to represent their learnings. Virtual meetings were arranged periodically for two sister schools to exchange ideas and share student works. The program lasted three school years from Sept. 2015 to May 2018.

In the first school year, there were two video conferences were held, and two exploratory activities were shared to the Chinese partner. Between September 2016 and May 2018, the learning space was successfully built and regular communication pathways have been set up for both sides. Five collaborative activities took place, and six video conferences were set for students to exchange their research progress and/or share their works. The teachers also took these chances to build rapport, enhance further understanding of each other's culture, and exchange thoughts and ideas for the purpose of making the project a better learning experience. The two sister schools that were involved in this program are quite representatives of both Windsor and Chongqing. On the Chinese side, the secondary school affiliated to the Southwest University started the 24 nature notes project in 2013, and it had been running successfully for two years before they developed a partnership with the Canadian school. According to the Chinese school's website (2014), this is a direct school under the China Ministry of Education and it is also a key school under the Chongqing City Department of Education. This means that this school has an excellent academic reputation in Chongqing, and the students in this school also have high academic standing. Like many other Chinese secondary schools, science was taught as separate subjects including biology, chemistry, and physics. This school has several science teachers to teach different subjects due to its large student population, and it traditionally provides an enriching environment for science education. The nature notes program was first launched in this school in 2013, and it was designed into a scientific-based activity since the designer of the nature notes project was the lead science teacher, and those who implemented the program were also from the science team. They became strong supporters in conducting this project, hence, it is no doubt that this project has been running smoothly from a scientific scope. The project had been continually running for two years before this Canadian-Chinese partnership was set up. The school runs from Grades 7 to 12, while the program was operated in Grades 7 to 8 and Grades 10 to 11. It became a part of the science curriculum and was implemented in different dimensions.

On the Canadian side, the public elementary school is located in a low socioeconomic community where students have diverse ethnic backgrounds. The school runs from Kindergarten to Grade 8. Like other Ontario schools, each class has a homeroom teacher who is responsible to teach a majority of the subjects. The teachers who were responsible for designing and implementing the nature notes reciprocal learning program in this school were not responsible for teaching science. This created minor obstacles when the school started to run the project in the 2015–16 school year. The teachers spent a year to explore some possible ways to integrate the project in their curriculum, but it did not go very well. However, both sides built a mechanism of information exchange during this year. Back-and-forth discussions were made in order to clarify the primary focus of the program. Two video conferences were conducted, and students shared information that they obtained about the nature as well as the work that they had done to represent their learnings. A breakthrough came in September 2016, two teachers took over the project, and they successfully integrated the project into the English literacy curriculum for Grade 7 and 8 levels. In order to reflect the project's scientific nature to the greatest extent, the teachers made the program into an inquiry-based learning activity. The goal that they set for this program was to enable students to learn the combined knowledge of English literacy and science. The students who participated in the nature notes program were mainly from two Grade 8 classes and one Grade 7/8 class. Most of the works related to the program were completed in school, but sometimes students were required to use their weekends and holidays to work on their projects in order to meet the set timeline. That work included outdoor observation, information searching, group discussions, filling out working packages, and designing a way to demonstrate their learnings.

Study Methods

To study the success of this program, the research team conducted a participatory qualitative research. The researchers were among the key members to coordinate the project and facilitate mutual communications. The project lasted for three school years and involved the Grade 7 and 8 students and two teachers from a Canadian sister school and the Grade 7, 8, 10, and 11 students and several science teachers from the Chinese sister school. Data collection started in 2015 on the first day of the project implementation while the majority of data was collected in the 2016–17 and 2017–18 school years as the research team conducted regular school visits as well as facilitated the research process in the last two years. The research team accumulated many kinds of data, which can be summarized into three types: documentary data, observation data, and interview data. These multiple types of data helped the researchers assuring the validity of this study and enhanced the transferability of findings (Burns, 1994) since these documentary data allowed for triangulation with interview data (Punch, 2005).

Documentary Data

The documentary data were collected by the research team during class visits and project meetings over a period of three years. They include project meeting minutes, video conference minutes, student observation reports of the nature, and photos that captured their working process and the accomplished products.

The meeting minutes are the records documenting teachers' thoughts and ideas that they exchanged with the partner school, and students' performance and experience over each work-sharing presentation through video conferences. Student observation reports consist the information that they searched for, the data that they collected, and the ideas that were generated during the process of working. Photos of the students' products were collected as valuable data as they led to the identification of changes in students' work. The changes include their perception over the project, which affected the topics that they chose and the ways that they used for presentation.

Observation Data

The researchers conducted class observations at different stages and components of student collaborative projects which include nature observation, inquiry question generation, discussions with partners, topic selection, and presentation design. The class observation field notes documented the teachers' instructional methods, their mutual understandings and interpretations toward the program and each other's cultures, and how such interpretations and understandings reflect on the program's implementation.

Interview Data

In order to collect participating teachers' and students' experiences and perceptions of the project, interviews and focus groups were conducted at the end of the second year and the beginning of the third year. Two Canadian teachers (both females) participated in one-on-one semi-structured interviews. Eleven Canadian students participated in two face-to-face focus groups. One focus group took place with six Grade 8 students, three females and three males. The other group had five student participants, all females. Six Chinese teachers (three males and three females) and three Grade 10 and three Grade 8 Chinese students (2 males and 4 females) were interviewed. These interviews and focus groups captured students' and teachers' perceptions and experiences toward various aspects of the program. Teachers recalled their experiences of implementing the program, the students' performances, the changes in their views toward each other's culture, the learning from sister school teachers, and their perception about such international collaboration. Students shared their experiences and feelings about how they explored their topics, how the project contributed to their science learning, what they learned from their sister school counterparts, and their suggestions for improvement of such international collaboration. In sum, the data gathered from school visits, video conferences, and interviews provided the researchers sound information to generate an understanding of how teachers and students from the two sister schools experienced the program and impacted each other through the program.

Data Analysis

An inductive approach (Berg, 2009; Miles et al., 2014) was adopted for data analysis. It includes three concurrent flows of action: data reduction, data display, and conclusion drawing and verification. The researchers first read through all text information several times, word by word, to label all key points and topics, and as well capture any possible deeper meaning from participants' narratives, researchers' field notes, meeting minutes, and photos. Cross-reading of all types of data was conducted for an accurate understanding of participants' learning experiences and achievements. The initial coding and memos were modified, classified, and assembled to several themes. Data analysis was cross-checked by the research team to assure the inter-rater reliability.

The Structure of the Book

Besides this introduction chapter, the book has six more chapters covering the origin of the international cross-cultural and collaborative educational program, the design and procedure, and its impact on participating teachers and students.

Chapter 2 introduces the Chinese traditional culture around the 24 solar terms and the farming culture in Canada around 12 constellations. According to the Chinese lunar calendar, the solar terms divide a year into 24 time periods. These 24 solar terms reflect the changes of seasons and climates, guide agricultural activities, and influence the basic necessities of thousands of households in ancient and modern China. The 12 constellations are a western concept developed based on the position of the Earth in its orbit. Similar to the 24 solar terms, the 12 constellations divide the year into 12 periods. These two concepts are coincidentally correlated.

Chapter 3 reports how the nature notes became an educational program in China and then took place simultaneously in two sister schools. The nature notes program is a framework allowing students to observe the natural transition at different times and generate their findings throughout the whole year of the study. This chapter covers two topics. The first topic is about the background information about the nature notes contest that took place in Chongqing where the Chinese school is located. The second topic is about how the nature notes program was run in the Chinese school before it became a collaborative program between two sister schools.

Chapter 4 uses student work samples to illustrate the achievements from the program. It describes how the nature notes program was implemented in two culturally different schools, how teachers respectively integrated it into their curriculum, how students learned, and how they presented their learning. The differences between Canadian and Chinese students are highlighted in terms of topics and methods to work on their projects and the ways to present their work.

Chapter 5 writes about the impact of the program on students. Based on interview data with participating students, the program's impact on students is explored from several aspects: understanding of nature, development of research skills, experience with reciprocal learning, choices of university majors and future careers, and so on.

Chapter 6 investigates teachers' perceptions of and experiences with the program. The participating teachers from the Canadian and Chinese sister schools were interviewed about their involvement in the program. During the interviews, teachers elaborated their feelings of being part of the program, the professional growth they experienced through the program, and their views of the effectiveness and limitations of this program.

Chapter 7 concludes the book by summarizing the impact of the program on participating students and teachers. The implications for future practices on cross-cultural learning programs are discussed.

Just as the nature notes reciprocal learning program could not take place without collaboration between university researchers, school teachers, and graduate assistants, this book is a product of collective efforts. Professor George Zhou from the University of Windsor and Professor Yuanrong Li from Southwest University were the program leaders who oversaw the design and implementation of the program. Mr. Jian Luo is the leader of the nature notes program in the Chinese sister school and served as the key resource person for the reciprocal learning program at the beginning stage. Zhou and Li reviewed, revised, and finalized the initial drafts of all chapters. The following persons participated in drafting the chapters: Xue Chen and Yanan Liao (Chap. 2); Shijing Guo, Sonia Ho, Qian Yang, Han Deng, and Jiang Zeng (Chap. 3); Fei Tao, Sonia Ho, Yangxue Ou, Ruihan Wang, and Yan Tian (Chap. 4); Sonia Ho, Jin Mu, Yiling Qi, Liujun Guo, and Shijie Deng (Chap. 5); Sonia Ho, Guihua Zhang, Yanhong Mao, Han Yang, and Xin Chen (Chap. 6); Shijing Guo, Banglong Jiang, and Sonia Ho (Chap. 7).

Chapter drafts were partially written in Chinese. Lin Li and Sonia Ho translated the Chinese writing into English. Gelsea Pizzuto conducted copy-editing of the manuscript.

We want to acknowledge school teachers from two sister schools: Haley Freedman and Carly Haggman (Prince Edward Elementary School); Xiaoqiang Ning, Li Chen, Jing Deng, Jie Song, and Xing Lü (Southwest University Affiliated Secondary School). The following graduate student assistants participated in the program in various roles: Peiyu Wang, Tian Liu, Junyi Yang, Yuan Sun, Zongrong Yu, Lizhi He, Xiao Shao, Chunlei Liu, Jing Tao, Jia Chen (University of Windsor); Pei Shi, Jianying Hu, Ting Song, Yuwen Tian, Kunhe Ma, Wei Ren, Xiangyi Wang, Jie Tao, Li Li. Chenyin Fu, Guoqing Zong, Mingyue Luo, Shue Wu, Qianwei Wang, Mingrui Chen, Xiaorong Fei, Yani Wu, Jiajia Deng, Jun Ma, Lianhua Zhang, Yanyue Tan, Yunzhi Xie, Nian Yi, Jia Wang, Pengfei Ruan, Pei Li, Jing Zhang, Xiaoliang Yang, Xue Wang, Jiao Tan, Tiancheng Lan, Shanshan Zhao, Kun Qian, Lijuan Yang, Xin Yi (Southwest University). Last but not least, the program owes many thanks to the administrative support from both sister schools.

References

Berg, B. L. (2009). *Qualitative research methods for the social sciences* (7th ed.). Pearson.

Boud, D. (1999). Situating academic development in professional work: Using peer learning. *International Journal for Academic Development, 4*(1), 3–10. https://doi.org/10.1080/1360144990040102

Burns, R. B. (1994). *Introduction to research methods.* Longman.

Carney, S. (2009). Negotiating policy in an age of globalization: Exploring educational "policyscapes" in Denmark, Nepal, and China. *Comparative Education Review, 53*(1), 63–88. https://doi.org/10.1086/593152

Connelly, F. M., & Xu, S. J. (2019). Reciprocal learning in the partnership project: From knowing to doing in comparative research models. *Teachers and Teaching, 25*(6), 627–646. https://doi.org/10.1080/13540602.2019.1601077

Cowen, R. (2006). Acting comparatively upon the educational world: Puzzles and possibilities. *Oxford Review of Education, 32*(5), 561–573. https://doi.org/10.1080/03054980600976155

Hofstein, A., & Rosenfeld, S. (1996). Bridging the gap between formal and informal science learning. *Studies in Science Education, 28*, 87–112.

Lewis, S., Sellar, S., & Lingard, B. (2016). PISA for schools: Topological rationality and new spaces of the OECD's global educational governance. *Comparative Education Review, 60*(1), 27–57.

Miles, M. B., Huberman, A. M., & Saldana, J. (2014). *Qualitative data analysis* (3rd ed.). Sage.

Punch, K. F. (2005). *Introduction to social research: Quantitative and qualitative approaches* (2nd ed.). Sage.

Zhou, G., Ho, S. W.-Y., Li, Y., Luo, M., Freedman, H., & Luo, J. (2019). Reciprocal learning between Canadian and Chinese schools through the 24 nature notes project. *Journal of Teaching and Learning. Special Issue: Reciprocal Learning., 13*(1), 7–24. https://doi.org/10.22329/jtl.v13i1.5988

CHAPTER 2

Following the Rhythm of Nature: Wisdom in the 24 Solar Terms and the 12 Constellations

Xue Chen, Yanan Liao, and Yuanrong Li

Chinese Traditional Culture

Chinese traditional culture is one of the most vital cultural systems in the world. For over 5,000 years, the system has been living and breathing with its profound and inclusive spirit, exerting a significant influence on the formation and development of the values and lifestyles of the people in China and neighboring countries. There are various interpretations of the Chinese traditional cultural system, but it is no doubt that Confucianism, Taoism (Daoism), and Buddhism constitute its core and fundamental

X. Chen • Y. Liao
Southwest University, Chongqing, ON, China

Y. Li (✉)
Southwest University, Chongqing, China

College of Teacher Education, Southwest University, Chongqing, China
e-mail: hxliyr@swu.edu.cn

G. Zhou et al. (eds.), *Science Education and International Cross-Cultural Reciprocal Learning*, Intercultural Reciprocal Learning in Chinese and Western Education,
https://doi.org/10.1007/978-3-031-17157-4_2

values. These ideas and values permeated the ancient Chinese political, economic, and legal systems. They were expressed through art, architecture, literature, and medicine, creating a distinctive cultural style different from the Western culture and other native cultures on different continents. Side by side with these cultures, they constitute a pluralistic coexistence of world cultures.

However, in modern history, Chinese people suffered an unprecedented setback and the resulting humiliation, prompting them to reflect on the contrast from glory to backwardness. Because of the overwhelming desire to learn from the West and to revitalize the nation, some thinkers, in a radical and either/or way, attributed the cause of China's backwardness in modern times to the traditional culture of emphasizing literature over science. As a result, Chinese traditional culture was turned into a synonym for "ignorance," a deep-rooted concept that any initiative to emphasize the traditional culture was seen as stubborn and obstinate (Huang, 2017; Luo, 1999, 2010; Wang, 2007).

Today, in the twenty-first century, with the development of China's economy and the strengthening of its national power, there is a growing acceptance of the belief that the traditional Chinese culture is not a hindrance to social development (Li, 2000; Liu, 2006, 2008; Zhang, 2007). Instead, economic development requires cultural willpower, and China's development and social stability require an understanding of the long history of the culture. At the same time, one of the important discoveries of the globalization process has been a new understanding of the relationship between a nation and the world. There is a growing recognition that the increased exchange between cultures worldwide has fostered mutual understanding, tolerance, and learning among cultures. Multiculturalism is the essence of a wonderfully diverse world of humanities. The exchange and interaction of different cultures do not make a particular culture disappear. However, it is a process where people of various cultural backgrounds understand and appreciate each other. Such interacting and learning processes constantly strengthen the uniqueness of their own cultures while evolving to be more international.

Science in Ancient China

As mentioned above, many people think that the traditional Chinese culture emphasizes literature over science. But this belief is biased. The history of traditional Chinese science and technology is also long-standing, including the development of astronomy, geology, agriculture, medicine, physics, chemistry, and museums, to name a few.

Taking astronomy as an example, the first part of the Confucian classic book *Shang Shu* (Book of History, High Official, Government Minister) is entitled *Scripture of Yao*. The section of *Scripture of Yao* tells the story of the famous ancient rulers, Yao and Shun, who appointed people by virtue, eliminated evil, loved the people, worked diligently, and gave their best (Liu, 2000). *Scripture of Yao* has a record of people using a jade balance to align the movement of seven stars. The jade balance is deemed as jade-based astronomical instruments; the seven stars refer to the sun, the moon, and the five other then-known stars. Here, it indicates that jade astronomical instruments were used to measure the moving patterns of the sun and the planets. These records have shown that astronomy was born at about the same time that agricultural science and medicine were created. Recognizing celestial phenomena and making calendars were already conscious actions, so it seems that ancient Chinese astronomy had already developed to a considerable extent around 2100 BC. Let us look at another classic book *Mr. Lü's Spring and Autumn (Annals)* (Zhang, 2012). This book series provides a comprehensive account of the ancient and modern events in heaven and earth. In its first 12 volumes, the book mainly discusses political, ethical, or philosophical matters, but the first chapter of each volume is about celestial signs and seasons.

Suppose the documents mentioned above can only be regarded as the science of the legendary era or the pre-Qin dynasty era, which is difficult to distinguish the truth from the falsehood. In that case, there is more solid evidence for scientific development in the later time in the history of China. In the Wei, Jin, and North and South Dynasties, there was a dispute over the calendar and a discussion of the Floating Heaven Theory. In the Sui and Tang Dynasties, discoveries were recorded in the book *Sui Shu* about the movement of the sun, the moon, and the five stars (Yi, 2016). In the Song and Yuan Dynasties, the origin of the universe was discussed. At the end of the Ming Dynasty, Chinese science started to be integrated into the world trend of scientific development. The missionaries brought new methods of arithmetic and mathematical knowledge to China. During the Qing Dynasty, the book *Illustrated Gazette of Overseas States* (Wei, 1999) introduced Copernicus' geocentric theory.

The 24 Solar Terms

The fundamental problem of ancient astronomy, in a nutshell, is to project the positions of the sun, the moon, and the five stars on the celestial sphere at a given time and place. It is true in both the East and West in ancient times. The various astrophysical problems that dominate modern astronomy did not exist in ancient times. Then we can perhaps reduce it to a calendrical problem. Many scholars in ancient and modern times mentioned that a calendar serves agriculture. If we want to talk about the relationship between a calendar and agriculture, we have to discuss the 24 Solar Terms.

The 24 Solar Terms are a unique traditional Chinese calendar covering everything from wind, rain, thunder, and lightning down to all living beings. Chinese laborers created this unique calendar during the long-term agricultural production practice by continuously observing the sun and celestial phenomena. After continuous exploration, analysis, and summary, the division of the Solar Terms gradually became scientific and meaningful (Fig. 2.1). The complete names of the 24 Solar Terms were first published in the book *Huainanzi—Astronomical Training* (Miscellany of Writing from the Western Han) (Mei, 2011). In the early Western Han Dynasty, some of the names were found in the pre-Qin texts. In the calendars handed down to us, when there is a list of the 24 Solar Terms, the "72 Hou" are often attached to each of them, such as in the Dayan Calendar (*Research Group on the History of Chinese Astronomy*, 1981). It is clear the origin of the 24 Solar Terms is related to the observation of the weather by the ancestors.

The 24 Solar Terms are the wisdom of the ancients to summarize the changes in the sky, climate, weather, agriculture, and seasonal changes in the year, which is the unique knowledge system of time and social practice of Chinese people, guiding the traditional agricultural production and people's daily life from ancient times to the present (Research Initiatives, n.d.).

The **Start of Spring** (Li Chun, 立春) is the day when the sun's celestial longitude is at 315 degrees between February 3 and 5 every year. The spring season begins to descend on this day, which is the beginning of the four seasons, so it is called the Start of Spring. Like many other festivals, this festival has many folk customs, such as celebratory rituals to welcome and celebrate spring with Cattle Beating or Biting Spring. China is a large agricultural country, and people have always attached importance to

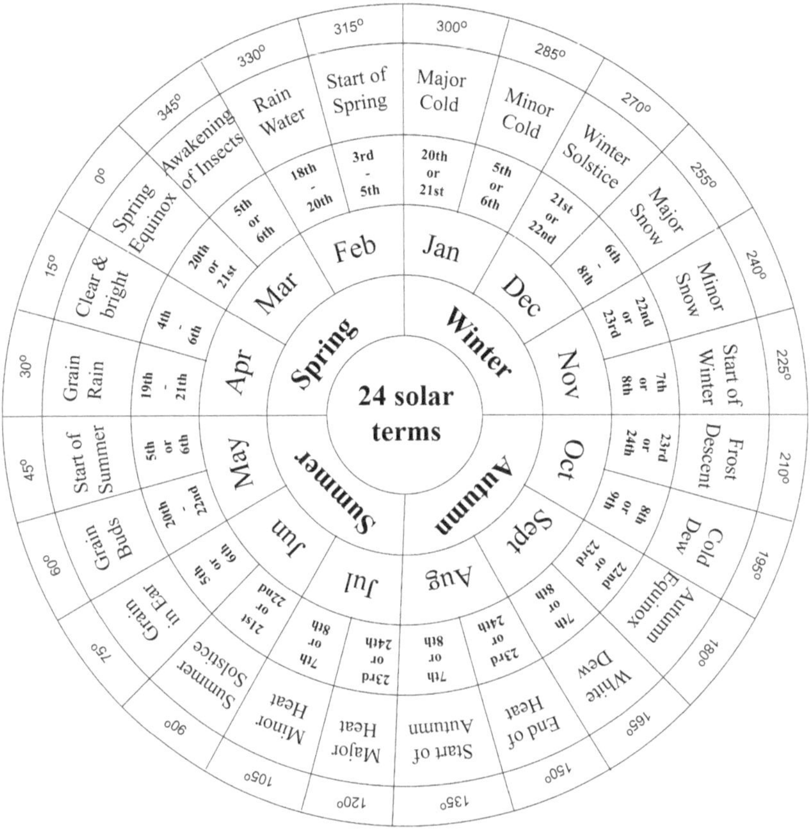

Fig. 2.1 Correspondence between the 24 Solar Terms and the Lunar Calendar

agricultural production and adhered to agricultural timing. Cattle symbolize farming, representing productivity, so Chinese people have welcomed spring with cattle since ancient times. People set up clay cattle and beat them with ropes to remind farmers that it was time to start thinking of planting. The so-called Biting Spring refers to the custom of eating spring plates, spring cakes, spring rolls, and chewing radish on the day of the Start of Spring. The word "bite" describes the many food customs of the season.

Rain Water (Yu Shui, 雨水) starts between February 18–20 every year when the sun reaches 330 degrees of the celestial longitude, meaning that the precipitation increases. Rainwater Festival is the second festival after the Start of Spring. Chinese people express the importance of spring rain using the statement that spring rain is as expensive as oil. There are some special activities on this day in many places, such as Pulling a Godfather (La Bao Bao) and Fortune-telling from Rice Color (Zhan Dao Se). La Bao Bao means to pull a godfather. In ancient times, people believed in fate; parents would ask God for their children's fortune, hoping to find out whether it would be easy to raise the children or not. Some parents might worry about their single child's death. Therefore, parents often found a godfather for their child, expecting the godfather's good fortune to shape the child. Fortune-telling from Rice Color is a popular custom in Southern China's rice production areas. It divines the harvest of the year's rice by frying glutinous rice. People put rice in a fire. If more rice is popped, the year will have a good harvest. The more pop rice, the better.

Awakening of Insects (Jing Zhe, 惊蛰) is celebrated yearly on March 5 or 6, when the sun reaches 345 degrees of the celestial longitude. It is the third of the 24 Solar Terms in the lunar calendar and marks the beginning of the mid-spring season. Before this date, animals find a hidden spot to sleep during the winter and do not drink or eat, which is called hibernation. At the festival of Awakening of Insects, the spring thunder wakes up the hibernating animals, from which the name Awakening of Insects originated. At the point of this festival, hibernating insects wake up, the weather becomes warmer, more spring thunder gradually appears, and most areas of China enter the spring farming season. For a long time, Chinese communities have developed many celebration events, such as Sacrificing White Tiger, Frying Insects, and Hitting the Villain. In the ancient agrarian societies, tigers were a common problem, so people sacrificed white tigers on the day of the Awakening of Insects for the sake of peace. Starting at the festival of Awakening of Insects, animals come out and wander around, and white tigers also come out to pray at this time. To pray for peace, people sacrifice the white tiger on this day. They believe white tigers will not eat humans once they are fed meat. At the same time, insects come out of the dirt and caves. They gradually explore the fields and the homes, harming crops or disrupting everyday life. Therefore, during the Awakening of Insects, there are different rituals to get rid of insects in different places. In addition, in ancient times, people would hold incense and mugwort to smoke the four walls of their homes to drive away snakes, insects, ants, rats, and moldy odors with the fragrance. As time went by, this tradition has gradually evolved into the custom of slapping the adversary and driving away bad luck, which is also the prototype of Villain Hitting.

Spring Equinox (Chun Fen, 春分) takes place on March 20 or 21 every year when the sun reaches 0 degrees of celestial longitude. Spring Equinox ranks fourth of the 24 Solar Terms. It takes place in the middle of the ninety days of spring. On this day, daytime and nighttime are at equal lengths. The term Spring Equinox is rich in connotations, and there are various events and activities, such as Worshipping Ancestry, Eating Spring Vegetables, and Standing-Up Eggs. As early as the Zhou Dynasty, there was a ritual for Spring Equinox, and this custom has been passed down through the generations since then. Moreover, Spring Equinox is when the spring thunder starts, everything revives, so it is the right time to eat spring vegetables. In the past, the people of the four eastern regions of Lingnan had a custom called Eating Spring Vegetables at Spring Equinox. In addition, legend has it that it is easier to stand up an egg on the day of Spring Equinox. According to historical records, the tradition of Standing-Up Eggs on Vernal Equinox originated 4000 years ago in China, where people celebrated the coming of spring.

Clear and Bright (Qing Ming, 清明) is the period between April 4 and 6 every year when the sun is located at 15 degrees of the celestial longitude. Clear and Bright means a clear sky and bright earth. In other words, when the wind of Clear and Bright arrives, everything is clean. The folklore activities of Clear and Bright are mainly commemorating ancestors, visiting graves, going on spring excursions, and planting trees. Through the Clear and Bright Festival, Chinese people remember their ancestors and give thanks to them, conducive to family harmony and social stability.

Grain Rain (Gu Yu, 谷雨), between April 19 and 21 of the Gregorian calendar, is the day when the sun reaches 30 degrees of the celestial longitude, which is generally the third month of the lunar calendar, also known as Month Chen or the month of the silkworm. The plan of a year lies in spring, and Grain Rain is the key period of crop growth, so the importance of the Grain Rain season is self-evident. Because of this reason, there are customs associated with Grain Rain, such as Removing Five Poisons and Enjoying Peonies. To avoid the harm of pests to crops and people, people go to the fields to exterminate the pests. People also post the Gu Yu Stickers, which are made of yellow paper with the images of divine chickens catching scorpions and celestial masters removing the five poisons to deter the five poisons. At the same time, the peony flowers are in full bloom at the time of Grain Rain, so they are also called Grain Rain flowers, and the custom of *Enjoying Peonies at Grain Rain* has been derived.

Start of Summer (Li Xia, 立夏) is on May 5 or 6 of each year on the Gregorian calendar, when the sun reaches 45 degrees of celestial longitude and indicates the start of summer. In Chinese folklore, many traditional customs have gradually been established to welcome the arrival of summer, such as Buying Saffron and Hanging Summer Eggs. At the beginning of summer, in southern Zhejiang, every household buys saffron and new tea for the following year. At the same time, the ailment Hanging Summer is a common condition in summer with symptoms of bloating and anorexia, fatigue, and weight loss. At noon on the day of the first summer, people boil eggs (without shell cracks), merge them in cold water for a few minutes and then put them in a long-woven wire mesh bag and hang them on the child's neck to pray for healthy growth in the summer.

Grain Buds (Xiao Man, 小满) takes place between May 20 and 22 when the sun reaches 60 degrees of celestial longitude. During this time in northern China, the seeds of wheat and other summer crops have begun to grow fully but not yet mature, roughly in a period of late milk ripening, so it is called Grain Buds. Grain Buds refelct the agricultural season, and the folk customs associated with this period are also connected with agricultural production, such as Worshipping the God of Vehicles, Busy Summer Meetings, and Grabbing Water to name a few. In the celebration of Worshiping the God of Vehicles, folks worship three types of farming-related machines: irrigation, weaving, and oil-squeezing machines. Farmer's life cannot be separated from these machines. At the same time, Busy Summer Meetings often take place. Their primary purpose is to exchange and purchase farming tools, buy and sell livestock and produces. The event of Water Grabbing is a traditional farming custom that spreads around Haining, Zhejiang Province, and is usually organized by the elders in a village. At the dawn of a determined date, the organizers summon all households to channel river water to the farmers' fields until the river dries out.

Grain in Ear (Mang Zhong, 芒种) takes place each year on June 5 or 6 of the solar calendar when the sun's celestial longitude is 75 degrees. Grain in Ear is the cut-off point for the timing of planting crops. The weather starts to heat up at this time of the year. The wheat becomes ripe and the summer planting starts. It is the busiest season. The associate customs include Sending the God of Flowers, Boiling Plums, and Playing Mud. At the time of Grain in Ear, flowers have started to wither. Thus, farmers believe the flower God has retired and holds a ceremony to worship the flower God and send the God back to his place; Meanwhile, farmers express their gratitude to the flower God and hope to meet him again in the coming year. The custom of boiling plums at the solar term has a long history, as early as the Xia Dynasty. As plums are sour and astringent, they are difficult to eat directly and must be processed before they can be eaten, so they are boiled. In addition, the Dong people in Liping County, Qiandongnan Autonomous Prefecture, Guizhou Province, have an annual mud fighting festival in which young men and women have fun around the time of planting seedlings.

Summer Solstice (Xia Zhi, 夏至) occurs annually on June 21 or 22 of the Gregorian calendar when the sun reaches 90 degrees of the celestial longitude. Summer Solstice has the longest daytime and the shortest night of the year, and it signals the start of hot days. The main customs of this solar term include eating noodles and body weighing. Eating noodles on Summer Solstice is very popular in many areas of China because new wheat has just been harvested. Eating noodles on this occasion implies tasting something fresh and enjoying the harvest. Folklore says, "winter solstice dumplings and summer solstice noodles," which point out the traditional foods on these two special days. Weighing on Summer Solstice day is to check the shape of the body: overweight or underweight. Farmers believe that weighing during this festival will help them survive the hot summer days to come and bring them good luck.

Minor Heat (Xiao Shu, 小暑) is the day when the sun reaches 105 degrees of the celestial longitude on July 7 or 8 of each year in the Gregorian calendar. The weather starts to become hot, but not yet the hottest. The main customs of the solar term are eating vermicelli and wearing jasmine. The weather starts to get hot after Minor Heat, so people eat cool or cold foods to get through the hot days. In Jiangsu and Shanghai, flower sellers in summer often tie the jasmine flowers with thin wire into balls and bands for women to wear.

Major Heat (Da Shu, 大暑) occurs annually on July 23 or 24 of the Gregorian calendar when the sun reaches 120 degrees of the celestial longitude. It is the hottest period of the year with the fastest growth of heat-tolerating crops. The main customs of Major Heat include drinking summer goat soup and eating magic grass. In the southeast of Shandong province, it is customary to drink goat soup on this day. In Guangdong, people have the custom of eating magic grass in the summer. The grass, also known as fairy grass, is an important dual-use plant for medicine and food and has a magical effect of cooling off the summer heat, so it is called magic grass.

Start of Autumn (Li Qiu, 立秋) occurs when the sun reaches 135 degrees of the celestial longitude on August 7 or 8 of each year on the Gregorian calendar. With the start of autumn, the summer heat goes away, and the cool weather gradually arrives. Crops will ripen soon. Chinese people attach great importance to the fall season, so there are many folk customs about this solar term. The Feudal Emperors often had an altar set up and welcomed autumn's arrival with their civil and military officials. The public carried the customs of putting on autumn fat and gnawing autumn in communities. In ancient times of China, a person's health was often judged only by the standard of fat under the skin, and it was popular among common folks to weigh a person on the day of the Start of Autumn to compare their weight with that in the summer to judge their health status. If they were thinner, they had to make up for it after the Start of Autumn by putting on autumn fat. The way to do this is oil-thick delicacies. In China, many places have the custom of Gnawing Autumn, also known as Biting Autumn, where people believe that eating something at the time of the Start of Autumn can prevent diseases from occurring.

End of Heat (Chu Shu, 处暑) occurs on August 23 or 24 of each year on the Gregorian calendar when the sun reaches 150 degrees of the celestial longitude. It is a festival that reflects the temperature change. The End of Heat means the hot weather is coming to an end. There are many festivals around the End of Heat, such as the Chinese Valentine's Day and Midyear Festival, to which common folks attach significant importance. During these festivals, the lozenge is flourishing. Women on the swinging orchid dressed in commoner skirts sing and pick up lozenges in the breeze. The custom of splashing water can be seen during the month from Major Heat to the End of Heat in Southern China. At the festival of splashing water, folks splash water on each other to have fun and cool down themselves.

White Dew (Bai Lu, 白露) comes on September 7 or 8 of each year on the Gregorian calendar, when the sun reaches 165 degrees of the celestial longitude. After White Dew, the night gets colder each day. Therefore, it is customary to eat food to replenish the body's energy and prepare for the cold winter. During White Dew, people will drink White Dew Tea and eat longan. In the city of Nanjing, the older adults there are very fond of drinking White Dew Tea when the tea has a unique sweet and clear flavor.

People in the Fuzhou area during White Dew must eat longan, and they believe longan can replenish their energy.

Autumn Equinox (Qiu Fen, 秋分) is on September 22 or 23 every year when the sun reaches 180 degrees of celestial longitude. On Autumn Equinox, the sun shines directly on the Earth's equator, and the 24 hours are equally divided into day and night with 12 hours for each. Autumn Equinox is also exactly at the middle of the 90 days of autumn, which equally divides the season. Chinese people celebrate the Autumn Equinox through such events as erecting eggs and sending autumn cattle. At Autumn Equinox, people choose fresh eggs that have just been laid for four or five days and are smooth and well-proportioned, and then gently make the eggs stand vertically up on a table. People go from door to door and deliver a picture of an autumn cow. The picture was drawn on red or yellow paper with a picture of a farmer plowing the field with a cow.

Cold Dew (Han Lu, 寒露) is on October 8 or 9 of each year on the Gregorian calendar when the sun reaches 195 degrees of the celestial longitude. Dew is a sign that the weather has turned cold, and the water vapor in the air is condensed to form dew at night. When the weather turns cold on Cold Dew, mountain climbing becomes a major custom everywhere.

Frost Descent (Shuang Jiang, 霜降) is on October 23 or 24 of each year on the Gregorian calendar when the sun is located at 210 degrees of the celestial longitude. Frost Descent is the last festival of autumn. The dew condenses into frost, and the colder weather can be felt everywhere. During this transition time from autumn to winter, Chinese local food customs are distinctive, as people believe that they should eat special food to supplement their bodies. Persimmons and apples are often among the dietary supplements. Merchants put chestnuts and persimmons together to pray for a profitable market. The combination of the pronunciation of Chinese names of chestnuts and persimmons makes a sound that means a favorable market with profit.

Start of Winter (Li Dong, 立冬) is set on November 7 or 8 of each year on the solar calendar when the sun reaches 225 degrees of the celestial longitude. It marks the beginning of winter and the arrival of cold weather. People attach great importance to the Start of Winter, the first winter festival. From ancient times to the present, there are rich and diverse customs, such as warming the fireplace, frying herbs, and eating lamb to name a few. On the day of the Start of Winter, people set up a stove and burn charcoal to have a warming party. People use various herbs, chrysanthemums, and honeysuckle to make a decoction for bathing. In addition, weak people begin to eat tonics. Mutton has become popular in the market since Chinese people see mutton as one kind of warm food. After eating warm food, people will feel warm inside their bodies. In contrast, some food is cold food, such as crab. People will feel cold inside after eating crab. Here, warm or cold does not refer to the temperature of the food. They are terms in Chinese medicine.

Minor Snow (Xiao Xue, 小雪) is on November 22 or 23 of the Gregorian calendar when the sun reaches 240 degrees of the celestial longitude. After Minor Snow, the temperature drops sharply, and the weather becomes very dry, making it a good time for Chinese farmers to make sausages and bacon.

Major Snow (Da Xue, 大雪) is between December 6 and 8 every year in the Gregorian calendar when the sun reaches 255 degrees of the celestial longitude. The northern part of China is often affected by cold air and heavy snowfall, which results in the snow accumulated on the ground. Major Snow is relative to Minor Snow. After Major Snow, the day gets even colder, and snowfall increases in frequency and volume. Skiing, skating, and snow-ice racing are some of the common activities people do.

Winter Solstice (Dong Zhi, 冬至) is December 21 or 22 of each year on the Gregorian calendar when the sun reaches 270 degrees of the celestial longitude. Winter Solstice has the shortest day and longest night. The weather starts to enter the cold period of the year from then on. Chinese folklore states that the winter solstice is like the New Year Festival and people celebrate the winter solstice with many customs. For example, people in the north slaughter sheep and eat dumplings and wontons on Winter Solstice. People in the south eat rice dumplings and long noodles on this day. In some regions, people worship the ancestors in this solar term.

Minor Cold (Xiao Han, 小寒) is on January 5 or 6 of each year when the sun reaches 285 degrees of the celestial longitude. The temperature will drop below zero once the Minor Cold is over, particularly in Northern China. The geographical span from the north to the south of China is large, thus forming various local customs. Around Minor Cold, people skate and ride donkeys to explore plum blossoms.

Major Cold (Da Han, 大寒) is on January 20 or 21 of each year when the sun reaches 300 degrees of the celestial longitude, which is the last one of the 24 Solar Terms and the coldest part of the year. As the old saying goes, *Minor Cold and Major Cold, freezing into a cluster*. This is the coldest period of the year. People all over China begin to get busy saying goodbye to the past year and welcome the new one. They sweep the dust and clean rooms, and prepare goods for the celebration of the New Year. Many important folk customs and festivals happen during this solar term, such as having a year-end meal and eating glutinous rice. The year-end festival originates from the custom of worshipping God of the Land, and it is a way to thank God for blessing the harvest for the farmers and the fortune for the business persons. Meanwhile, the boss of each company would reward the staff at the company, his home, or a restaurant as a token of appreciation for their hard work throughout the year. In a large area of southern China, eating glutinous rice in Major Cold is a custom. Although the custom sounds simple, it contains the life experience accumulated in the past. Since the weather is freezing cold on Major Cold and glutinous rice is high in calories, eating glutinous rice has an excellent effect in helping people fight the cold.

The 24 Solar Terms have had a profound influence on Chinese agricultural production. With the help of the Solar Terms, farmers divide the year into several stages in terms of the agricultural cycle: plowing, fertilizing, irrigating, harvesting, and storing. The 24 Solar Terms have also had a great influence on the living habits of Chinese people. In addition to the various folkloric activities practiced at different solar terms described above, the 24 Solar Terms have influenced Chinese people's diet and health habits. In different terms, people carry out different fitness exercises or adjust their diet to keep their bodies healthy. For example, in the spring, the days become longer and warmer. The body's blood metabolism is strong. People should eat less acidic food and they should protect and nourish the liver. During a snowy season, the temperature continuously drops and the occurrences of asthma are high. This is the time people should take the tonic. They should also take appropriate exercise to enhance physical fitness (Yu & Lao, 2016).

The 24 Solar Terms Were Successfully Applied for the Heritage

With the development of modern technology, human beings can forecast the weather with scientific data. The function of the 24 Solar Terms has diminished in guiding agricultural affairs. However, the 24 Solar Terms have produced very few errors. Despite extreme weather phenomena, the cultivation of crops on farmland nowadays has not deviated from the original seasonal patterns (Bi, 2016). On November 30, 2016, the *24 Solar Terms—A System of Time Knowledge and Its Practice Developed by the Chinese through Observing the Annual Movement of the Sun* was officially inscribed on the UNESCO Representative List of the Intangible Cultural Heritage of Humanity. The presence on the UNESCO list reflects the importance that the international community attaches to safeguarding intangible cultural heritage in the form of traditional knowledge and practices. It also implies the recognition of this outstanding cultural heritage and the trust in Chinese responsibility for its safeguarding.

However, at present, the heritage and protection of the 24 Solar Terms are facing challenges. The young generations are beginning to ignore the cultural heritage associated with the solar terms, and its continued traditions are declining in many places (Liu, 2017). Society should actively promote the inheritance and protection of the 24 Solar Terms in order to avoid losing its cultural significance and tradition. Promotion of the 24 Solar Terms in the community and education system should occur so that customs and cultural connotations contained in the solar terms can continue to be passed on.

The 12 Constellations

These 24 Solar Terms reflect the changes in seasons and climates, guide agricultural activity, and influence the basic necessities of thousands of households in China. The Solar Terms divide a year into 24 time periods based on the sun's position in a year. According to the Chinese Lunar calendar, each solar term falls on a different date. The 12 constellations are a western concept developed based on the position of the Earth in its orbit. Similar to the 24 Solar Terms, the 12 constellations divide the year into 12 periods. These two concepts are coincidentally correlated as shown in Fig. 2.2. For example, March 20 to March 21 means the Spring Equinox in China, while it corresponds to the beginning of Aries in Canada.

The 24 Solar Terms and the 12 constellations are two representatives of world traditional cultures. It is valuable for the young generation to be aware of these cultural heritages and generate cross-cultural understanding. The nature notes program reported in this book took the traditional culture as an anchor. Two sister schools from Canada and China went to nature at chosen solar terms for their observation-based inquiry on plants,

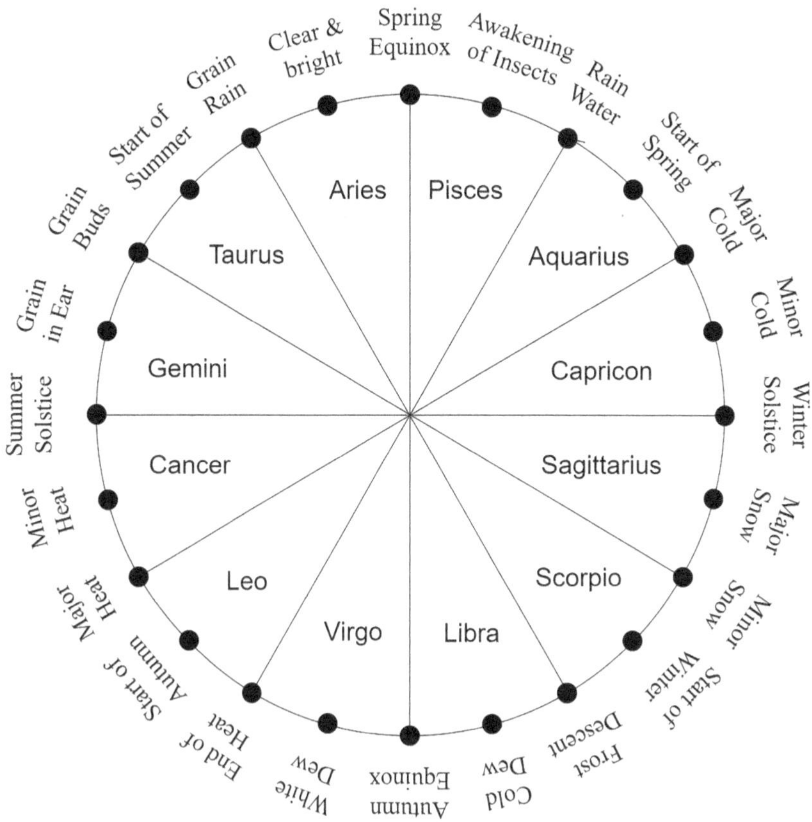

Fig. 2.2 Correspondence between the 24 Solar Terms and 12 Constellations

animals, geography, and so on. The implementation of the program and its impacts on participating teachers and students will be covered in the book.

References

Bi, X. L. (2016). *National cultural genes are the soul and roots of the Chinese dream.* http://culture.people.com.cn/n1/2016/1206/c22219-28927357.html

Huang, K. W. (2017). From "civilizational" discourse to "cultural" discourse: An important turn in Chinese thought in the late Qing and early Ming dynasties.

Journal of Nanjing University (Philosophy - Humanities - Social Sciences), 1, 68–78.
Li, Z. H. (2000). *Talking about Chinese and Western culture and art at the turn of the century Studies in Literature and Art, 2*, 24–31.
Liu, Z. D. (2000). Shang Shu - Yao Dian: An ancient Nuo opera "script". *Folk Art, 3*, 143–156.
Liu, X. Y. (2006). Modern neo-Confucianism's reflection on the anti-tradition of the May 4 movement: The case of Xu Fuguan. *Journal of Huazhong University of Science and Technology (Social Science Edition), 2*, 1–6.
Liu, S. X. (2008). The trajectory of modern neo-Confucianism. *Journal of Hangzhou Normal College (Social Science Edition), 1*, 1–6.
Liu, Z. D. (2017). The history of the 24 solar terms system and its modern transmission. *Cultural Heritage, 2*, 12–14. 157–158.
Luo, C. T. (1999). The division of the west: International storm clouds and the evolution of Chinese thought before and after May 4. *Chinese Social Sciences, 3*, 20–35.
Luo, C. T. (2010). He is always himself: The life and heart of Chen Duxiu. *Journal of Sichuan University (Philosophy and Social Science Edition), 5*, 48–60.
Mei, J. (2011). The evolution of the words of the upper ancient festivals and the formation of the names of the 24 festivals. *Journal of Huaihua College, 3*, 9–11.
Research Group on the History of Chinese Astronomy. (1981). *History of Chinese astronomy*. Science Press.
Research Initiatives. (n.d.). *The 24 solar terms website*. Retrieved January 13, 2021, from http://www.1mag.cn/index.htm
Wang, Q. S. (2007). How the new culture was "in motion": The new youth as a point of view. *Studies in Modern History, 1*, 21–40.
Wei, Y. (1999). *Hai guo tu zhi*. Zhongzhou Ancient Books Publishing House.
Yi, F. S. (2016). An erratum to the book of Jin, the book of sui, and the history of song on astronomy. *Studies in Chinese History, 2*, 62.
Yu, S. C., & Lao, S. (2016). *The book of time: Yu Shicun on the 24 solar terms*. China Friendship Publishing Company.
Zhang, R. L. (2007). Handan learns from the past, but loses his old steps: A discussion on the problem of "reverse geometry" in Chinese philosophical research. *Journal of Nanjing University(Philosophy. Humanities. Social Science Edition), 4*, 60–76.
Zhang, Z. (2012). A Comparison of the Agrarian ideas in Lu's spring and autumn: Shangnong and four other books and the agricultural records. *Chinese Agricultural History, 3*, 122–131.

CHAPTER 3

Same Seasons, Different Campuses: Setting off the Journey of Nature Notes Program

Shijing Guo, Sonia Ho, Qian Yang, Han Deng, Jiang Zeng, George Zhou, and Yuanrong Li

Nature Notes Contest

Contest Background

In recent years, there have been rapid developments in science and technology. Society increasingly requires the public to possess scientific literacy. Unfortunately, a number of studies have shown that young people have no interest in science and that they do not want to be scientists when they grow up (Yu et al., 2010). Making young people love science and

S. Guo • Q. Yang • H. Deng • J. Zeng • Y. Li
Southwest University, Chongqing, China
e-mail: hxliyr@swu.edu.cn

S. Ho • G. Zhou (✉)
University of Windsor, Windsor, ON, Canada
e-mail: gzhou@uwindsor.ca

G. Zhou et al. (eds.), *Science Education and International Cross-Cultural Reciprocal Learning*, Intercultural Reciprocal Learning in Chinese and Western Education,
https://doi.org/10.1007/978-3-031-17157-4_3

improve their scientific literacy has become one of the most fundamental challenges for education. At the same time, studies have shown that there is a weakening connection between today's youth and the natural world. Youth lack an understanding of the nature and show difficulty in appreciating that humans are part of the nature. They ignore the inseparable relationship between these two and exhibit a series of behavioural and psychological problems, which has been termed "Nature Deficit Disorder" (Li et al., 2017). The concept of "Nature Deficit Disorder" was first introduced by Richard Louv, an American journalist and child welfare advocate, who coined the term in his book entitled *The Last Child in the Woods.* Louv pointed out that today's children are showing an increase in obesity and attention disorders, which he identified as having much to do with the lack of contact with nature (Louv, 2010). In order to address such issues, China has launched a series of thematic activities in school education that organically combine nature and ecological education through the development of scientific literacy. The "Dream Classroom Nature Notes Competition" was launched jointly by a number of organizations: the Publicity Department of Chongqing Municipal Party Committee; the Municipal Education Commission; the Municipal Environment Bureau; the Municipal Science Association; and the Municipal Science Committee. The competition aims to:

- Encourage children and youth to go into nature and participate in scientific observation, investigation and research,
- Cultivate students' awareness and ability of scientific investigation,
- Feel the connection between human beings and nature,
- Experience the wisdom and beauty of nature,
- Appreciate life and be grateful to nature, and
- Enhance ecological civilization awareness and behaviour. (Youth Pioneer News, n.d.)

The first Nature Notes Contest was launched in 2014 in Chongqing. It offered a platform for all young people to genuinely participate in science and experience nature. In the past, various youth competitions, such as robotics, artificial intelligence or 3D printing competitions, had only been aimed at a small number of students. Most students could not join in because many schools had difficulties in providing the necessary computer equipment. At the current social context, these competitions were more specialized and required a lot of time and effort, which was challenging for ordinary children to master. In other words, it was unrealistic to promote these competitions to all young children. In contrast, the Nature Notes Contests engage all cities, families, schools, and children to participate since they do not require expensive equipment and offer a wide range of topics students can choose to work on. The contest takes the traditional Chinese 24 Solar Terms as the conceptual nodes and provides the public and students with nature education activities with well-designed themes, such as "Cherishing the Blessing of Farming," "Saving the Environment," and "Thanksgiving for the Nature." The Nature Notes Contest regards nature as a unique education space or platform. They encourage and guide the general public, especially young people and families, to go into nature to observe and record its beauty. By participating in the Nature Notes contest, students can discover the beauty around them, draw this natural beauty, and express their appreciation for it. When caring for every plant and tree around them, participants develop a love for their hometown and motherland and identify ways to establish relationships between humans and nature, between people and people, and between people and society.

Contest Themes

Since the first Nature Notes Contest in 2014, the contest has been held on six occasions, each one with a different theme. The theme for the first contest was "Appreciating Life, Giving Thanks to Nature." This promoted the ecological concept of respecting, following , and protecting nature. It encouraged the public to love nature and cherish the blessings we receive from it. To participate in this contest, one student observed the animals in a pond on campus and found red-eared tortoises (Fig. 3.1). He learned that people often treat red-eared tortoises as ornamental animals, which are invasive species in Chongqing. He commented that if pet owners no longer want to keep their red-eared tortoises, they should not release the pets into the Yangtze River. The pet tortoises should be released in a pond to avoid a larger number of reproductions. It is important to protect biodiversity.

One member of the Nature Note Contest judge committee commented: The observer found a turtle in the pool. After searching the literature, he learned that the red-eared leatherback turtle belongs to the family Zetidae. It is the most widely reared semi-aquatic reptile and is an invasive alien species in Chongqing city.

Fig. 3.1 Invasive Species Invade Campus. Note: An Entry of the First Chongqing Nature Note Contest (Author: Jin Liu, Southwest University Affiliated Secondary School. Mentor Teacher: Jian Luo)

The second contest advocated the values of *loving things and knowing grace, saving and cherishing blessings*. It encourages the public, especially young people and families, to go outdoors, observe and record nature, and record the beauty of nature through simple text and graphic descriptions. In one entry, a primary school student ZXJ observed the growth of wheat during the solar term Gain Buds (Xiaoman) (Fig. 3.2). In the fourth month of the lunar calendar (May in the calendar), the seeds of summer crops in northern China, such as wheat, begin to grow grain-bearing ears, although they are not yet fully ripe. This period of the year is called

Fig. 3.2 Little fullness, Little fullness, Wheat Gradually Becoming Full. Note: The first prize winner of the second Chongqing Nature Notes Contest (Author: Xijun Zhou, Chongqing High-tech Zone No. 1 Experimental Primary School. Mentor teacher: Haiyan Tang)

Xiaoman, which means "little fullness" in Chinese (Cheng, 1985). As ZXJ recorded in his nature notes: "The wheat grain is getting full!". They also expressed that "I saw the vigorous life force of wheat!"

One member of the contest judge committee commented: nature observation makes us understand that food is not easy to come and life is not easy to grow. No planting, no harvesting. We should cherish and be thankful for what we get through the virtue of examining ourselves.

The theme of the third Nature Notes contest was *Cherishing blessings, being grateful, loving nature.* This theme was intended to encourage parents to slow down from their busy work and for children to look up from the electronic screen. Parents and children should go out of their homes and schools and learn to appreciate the greatness of life and the wisdom of life through communication with plants and trees. They can learn to love nature, be friendly and grateful with all things, and use the nature resource sparingly, and cherish the blessings from nature. A primary school student XXR recorded a few moments of the growth journey of daffodils from seed bulbs to new leaves and then into bloom (Fig. 3.3). In her work, each vivid and beautiful painting was marked with a small paragraph of warm words, recording her feelings and emotions when she was making the painting. At the closing ceremony speech, XXR remarked: "I've loved observing nature since I was a child. When I walk along the road, I often stop and observe the things that are interesting to me." She uses the writing of nature notes as one way to relieve the stress of studying. She feels increasingly relaxed by getting close to nature.

One member of the contest judge committee comments: the young author recorded several fragments of the growth process of daffodils in the form of a diary, allowing us to understand the process of daffodils from the seed ball to the growth of new leaves and then to the blooming of flowers. In spring, daffodils are growing happily, and the young students who love daffodils are also growing happily. What a delightful picture!

The theme of the fourth contest was *Cherishing blessings, being kind, loving nature.* This contests was meant to advocate the values of loving things, being grateful, saving the resources, and cherishing blessings. The contest encouraged young people to observe the nature around them, discover the beauty around them, and express what they have learned. Through caring for the plants and trees around them, they can develop a love for their hometown and motherland and realize how to improve relations between people and nature, people and people, and people and society. A primary school student QHY discovered that onion flowers have

Fig. 3.3 Narcissus Has Spring. Note: The second prize Winner of the third Chongqing Nature Notes Contest (Author: Xie Xinrui, Huayan Primary School, Jiulongpo District. Mentor teacher: Luo Yi)

different shapes when they are observed from different angles (Fig. 3.4). She sighed and said that "there is no longer a cute little flower that looks exactly the same."

Zhimu Lin, a senior member of the Taiwan Association of Wild Nature Protection, served as a judge on the contest judge committee. He

Fig. 3.4 Onion Flower. Note: Award winner at the fourth Chongqing Nature Notes Contest (Author: Qian Hongying, Chengjiang Primary School, Beibei District. Mentor teacher: Zhou Lijuan)

commented: I think Hong Ying must have been attracted by the tiny onion flowers. She might never think that the onions that she usually eats would produce such flowers. Hong Ying is very careful to describe the flowers from different angles and present them naturally on the canvas and draw each small flower with care and patience. Such an attitude is admirable!

The theme of the fifth contest was *Feeling the City of Mountains and Waters, Painting the Beautiful Land*. This was intended to encourage the public, especially young students and families, to record those things, events, and people observed in nature in the form of handwriting, hand-drawing, or oral narration. Participants were encouraged to record the enlightenment and inspiration received from nature, to learn to appreciate nature and respect life, to learn to live in harmony with nature, to love nature, and to cherish the blessings from nature. Every year between Grain in Ear and Minor Heat, the warm and humid Pacific Ocean airflow carried on the southeast monsoon passes through the middle and lower reaches of the Yangtze River, Taiwan, south-central Japan and southern Korea. It causes continuous rainy weather in these areas. This time of period is called "Plum Rain Season" because it is the season plums are ripening in the south of the Yangtze River (Zhang, 2012). During the plum rain season, DHY, a secondary school student, traveled to Jinfo Mountain and observed a peculiar natural landscape: cloud fall (Fig. 3.5). Such an observation leads him to an inquiry of the formation of cloud fall.

Dr. Dongli Rui, one of the early advocators of the nature notes program in China, served on the contest judge committee. She commented: Nature notes, as a multidisciplinary and integrated inquiry-based curriculum for elementary and secondary school education, is receiving more and more attention from the education community. This nature notes piece is a vivid demonstration of the achievements of this curriculum. The author has clear logical thinking, uses proper methods, and attends to details in the inquiry process. His language is well organized and flows smoothly. It would be even better if the author could further improve his scientific drawing!

The theme of the sixth contest was *Taking Notes on the City of Water and Mountain, Writing about the Beauty of the Motherland*. From their communication with grass and trees, participants will understand the wisdom of life and know how to love things and appreciate the blessings of nature. While they walk in nature and record their observations, they will better remember their hometown and bless the country. Major Heat is the time of the year when the day has extended periods of sunshine and high temperature. The following proverb speaks about the behaviours of the snakes and other natural indicators at this season which single the weather:

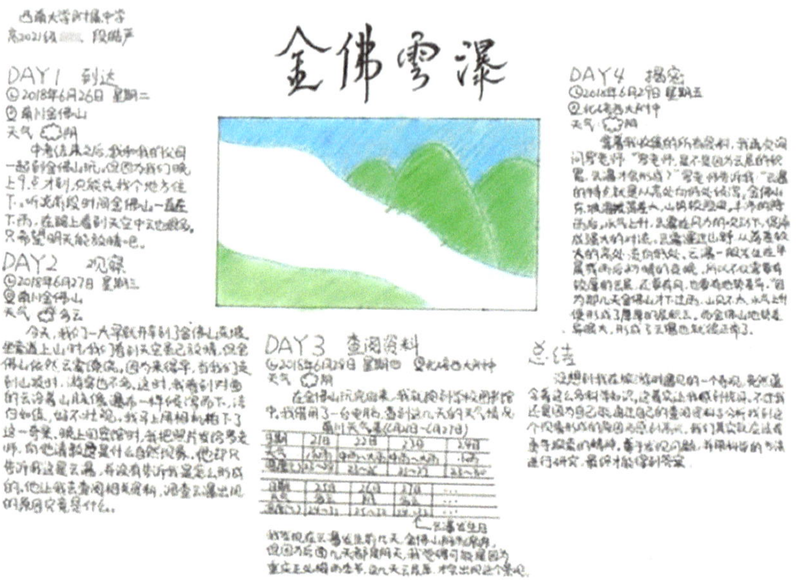

Fig. 3.5 Golden Buddha Cloud Fall. Note: The first prize winner of the 5th Chongqing Nature Notes Competition (Author: Duan Haoyan, Southwest University affiliated Secondary school. Mentor teacher: Jian Luo)

> *Swallows fly low and snakes cross the road,*
> *Chicken stays up late and toads cry.*
> *Salt pot gets damp and tobacco leave become wet,*
> *The heavy rain is coming soon.* (Dai & Jin, 2007)

During the study field trip at Major Heat, a primary school student SLH discovered a very special-looking snake: the gray-bellied green snake (Fig. 3.6). Her nature note project explored the question: Why is the gray-bellied green snake not green?

Hui Ou-Yang, the director of the Chongqing Nature Museum served on the contest judge committee. He commented: within the boundary of one sheet of paper, the content is rich. It includes the scientific investigation process of the gray-bellied green snake, the information on the distribution of the species across China, and safety tips for contacting with it. In addition, there is a comparison of the differences with the bamboo leaf green. Even more remarkable is that the author carefully chose the photos

Fig. 3.6 Encounter with Green and Black Eyebrows. Note. The first prize Winner of the 6th Chongqing Nature Notes Contest (Author: Shi-Luo Huiting, Southwest University Affiliated Primary School. Mentor teacher: Jian Luo and Yanqi Li)

and vividly recorded the whole study process. The project layout is colorful, vivid, attractive, and it complements the content.

Contest Participants

The target participant population of the Chongqing Nature Notes Contest has changed little since it first started. It is open to the public, particularly those young children aged 4–18. However, the grouping of participants has somewhat changed each year. For example, the 6th Nature Notes Contest participants were divided into three categories: family (children in kindergarten and their parents); primary school students; and secondary school students.

Contest Implementation

The Nature Notes Contest runs from May until November of each year. It consists of three phases: call for entries, review of entries, and award ceremony.

Call for Entries The contest theme is announced at the end of the previous year contest. Participants can start to submit their works in May. They mail their works to the local contest organizers including local schools and other educational communities. They can also submit their works through an online application or the social media platform (WeChat).

Review of Entries The submission review process goes through the following three stages: preliminary competition, second round competition, and final competition. First, each participating school or local educational community organizes its own preliminary review of submissions and recommends about 20% of its entries to the local contest committee. The local contest committee, supervised by the county or district government, will conduct the second-round competition and recommend 30% of its entries to the contest committee. Finally, in the middle of October, the contest judge committee will conduct the final evaluation. The winners will be announced on the contest's WeChat platform and its official webpage.

Award Ceremony The award ceremony takes place in the middle of November and award-winning projects will be announced at the ceremony. The award winners are invited to give a brief speech and receive a certificate.

In the process of the Nature Note Contest, a series of activities will be organized and carried out. The contest organizers will organize professional development workshops for nature education teachers. The large and small scale of symposiums or salons on nature education are held for idea sharing and inquiry. The organizers also set up local partners who help promote the contest program and pick up young advocators for nature education. Some nature locations are selected as the bases for nature education. Nature exhibitions (mobile zoo) are held in schools and local communities for children to learn to safely contact small animals such as insects, fogs, and snakes. Student works are compiled into videos for sharing with the public.

The Format of Works

Student entries in Chongqing Nature Notes Contest are presented on one piece of A4-sized (29.7 cm × 21 cm) paper. There are certain reasons for choosing A4 paper. First, A4 paper is commonly used and widely available. Secondly, the standard size of works is easy to compile into collections. Thirdly, such a size of work presentation reduces students' anxiety to work on the project. Students often think they can finish a project on such a small size of paper.

Evaluation

The Nature Notes Contest sets five criteria for evaluating works: authentic topics; careful observation and detailed text description; thinking and inspiration; interesting discovery; and presentation design. These five criteria are assigned with different weights (Table 3.1). The evaluation index focuses on scientific knowledge, skills and methods, inspiration, and the idea of environmental protection. It emphasizes the extension and inference from the experience with nature and encourages the embodiment of graphics and text.

Authentic Topics Students' work must be based on their own observations, reporting on real discoveries in daily life and in nature, rather than just collecting material found on the internet. In other words, the work should be original and authentic. The work should also be scientific.

Table 3.1 Nature Notes Contest Evaluation Criteria Index

1.	Authentic topics	**30** points
2.	Careful observation and detailed text description	**10** points
3.	Unique reflection and inspiration	**20** points
4.	Interesting discoveries	**20** points
5.	Beautiful drawings and layout design	**20** points

Students should describe and interpret scientifically what they have observed and reflected about nature. Students can ensure a correct understanding of the observed phenomenon through consulting their mentor teachers and referring to other resources. Given the significance of this criterion, the judge committee assigns it a weight of 30 points.

Careful Observations and Detailed Written Descriptions Contest participants are required to illustrate their work using graphs and text to demonstrate their careful observation of nature. This sets up a requirement in terms of participants' ability in writing and drawing. Since participants are often elementary and middle-school students with quite a range of age differences, participants' abilities using language and graphs vary a great deal. To focus on the aspects of scientific observation and reflection in the project and minimize the impact of participants' language ability, the judge committee decides to weight this criterion by only 10 points.

Unique Reflection and Inspiration Participants' works are required to reflect their own thinking and reflection. Participants are encouraged to incorporate their own thinking and reflection into their observations of nature and gain a sense of beauty of nature, an enriched understanding of nature, and a philosophical enlightenment from nature. This criterion is assigned 20 points.

Interesting Discoveries Students are asked to record interesting discoveries in their work. For example, a Western University-affiliated secondary school student discovered that she could identify the females and males of red-spotted toads by their acoustic spectra and recorded this discovery in her nature notes project. This criterion is assigned a weight of 20 points.

Beautiful Drawings and Layout Design Participants are required to have a beautiful layout design for their work. But the Nature Notes Contest is not a competition in calligraphy or painting. It sets up a reasonable objective in layout design, embodiment between text and graph, and neat writing. Participants are not expected to demonstrate stunning drawing work and beautiful writing, but only develop aesthetic sensibilities within their own age abilities. This criterion is assigned a value of 20 points.

The contest judge committee includes experts and scholars from various fields, such as NGO directors, university professors, high school teachers, and company CEOs. The committee members review participants' works against the scoring criteria. Each committee member recommends outstanding works to the whole committee. The committee collaboratively decides the score of these nominated works. There is also a one-vote rejection system, and works with obvious scientific errors will be rejected by one vote.

Nature Notes Enter Campus

Background of Nature Notes Program Development

In January 2000, the Full-time General Senior High School Curriculum Plan (Trial Revision) was published, stipulating that Research-Based Study is a compulsory comprehensive practical activity-based course. This was the first time that the concept of the research-based study was introduced and clarified by a document of the Chinese Ministry of Education (Ministry of Education, 2000). In April, the Guidelines for Implementing Research-based Study in General High Schools (for Trial Implementation) was disseminated, emphasizing the importance of research-based study in cultivating students' innovative spirit and practical ability (Ministry of Education, 2001b). In June, the Ministry of Education stipulated, in the Outline of Basic Education Curriculum Reform (for Trial Implementation), that research-based study and comprehensive practical activities are core components of the curriculum to be incorporated into the structure of the basic education curriculum, and began a trial run in September in general high schools in ten provinces and cities (Ministry of Education, 2001a). In May 2010, China's State Council Executive Meeting considered and adopted the Outline of the National Medium and Long-term Education Reform and Development Plan (2010–2020), which includes "actively carrying out a research-based study" as one of the key elements to comprehensively improve the quality of education for general high school students" (Ministry of Education, 2010). In 2017, the 19th Communist

Party Congress proposed that "we should implement the Party's education policy, develop quality education, and cultivate socialist builders and successors who are well-rounded in moral, intellectual, physical, and aesthetic" (Xu, 2017). Based on the background of the above national policy, the Southwest University Affiliated Secondary School, one of China's most famous high schools, explores how to carry out research-based studies more effectively based on the existing foundation.

Subsequently, Nature Notes were created as a vehicle for research-based learning. One of the significant features of Nature Notes is to take observational records of nature by drawing. When participants draw the observed natural phenomena with their own hands, stroke by stroke, they can better convey the natural subtleties and authenticity that daily excursions and photographs cannot convey. In addition, students can add necessary text descriptions next to their drawings, such as the name, mood, and surroundings of the recorded object, the time, place, and weather of the recorded object, the mood and feelings of the observer, and so on. As Leslie and Ross, authors of *Notes on Nature*, say, "many of us have been exploring deeper ways of being in nature, including learning about its forms, protecting the 'inhabitants' who 'live on borrowed time in the natural world, or meditating on the true meaning of the continuity of life." Taking nature notes is "a relatively simple way for people to reconnect with nature," "you just have to spend some time walking around nature, looking at it, and then reflecting on it in a journal" (Leslie & Ross, 2008).

The nature notes program at Southwestern University Affiliated Secondary School (SWUAHS) is designed for a wide range of students, including middle school students (Grades 7–9) and the first two-year high school students (Grades 10–11). Since 2014, the nature notes program has continued for nearly eight years, forming a complete system that has become influential in Chongqing and in China. The success of the nature notes program can be attributed to the following advantages enjoyed by the school.

Beibei District of Chongqing has a Unique Richness of Species and a Complex Ecological Environment Jinyun Mountain National Nature Reserve is a typical subtropical evergreen broad-leaved forest area and a gene pool of plant species in the middle and upper reaches of the Yangtze River. It is a valuable base for forest research, ecological teaching and environmental education. Located at the foot of Jinyun Mountain in Beibei, SWUHS has established a good cooperative relationship with Jinyun Mountain National

Nature Reserve and has taken the advantage of the rich curriculum resources of the reserve. Many students have conducted their nature notes projects in this area. In addition, the school campus is covered by many different plants and provides a platform for students to observe nature. The fortress in the annex is home to many small animals, attracting students to observe animal lives and produce many nature notes. The famous river-front flower windows on campus have also become a place where students can observe nature.

The School Has a Tradition and Atmosphere for Environmental and Sustainable Development Education As one of the first national "green schools," SWUHS often participates in (or holds) environmental education activities and competitions geared toward sustainable development. For example, with collaboration from Jinyun Mountain National Nature Reserve Administration and Chongqing Nature Public Welfare Development Center, it held the Walk in Jinyun Flying World—Campus Bird Watching Activity in April of 2016. The school also organized students to participate in the Silk Road Challenge – 2016 Youth Water Conservation Design Competition, Global Nature Day – 2015 Youth Natural Science Knowledge Challenge, and the 10th National Water Technology Invention Competition for Secondary School Students. Since 2000, the school's research-based learning activities under the theme of Education for Environment and Sustainable Development have been an essential pillar of the nature notes program.

Teachers Actively Promote Nature Notes on Campus SWUHS has formed a team of teachers who are dedicated to the research-based study curriculum with the theme of environmental and sustainable development education for a long time. Among them, Mr. Jian Luo is one of the founders of the Nature Notes Contest and the leader of the Nature Notes Program at SWUHS. With school administrative support, Mr. Luo and his team have been actively promoting the implementation of the nature notes program since 2014. On September 13, 2014, more than 20 teachers and students from SWUHS participated in the 2014 China National Joyful Coast Nature Notes Event organized by the Publicity and Education Center of the State Ministry of Environmental Protection with collaboration from the Publicity and Education Center of Chongqing Municipal Environmental Protection Bureau, SWUHS, and Chongqing Jinyun Mountain National Nature Reserve. The theme of this event was for middle and high school students to experience nature and take notes on Jinyun Mountain. The event participants observed and recorded the plants and climate of Jinyun Mountain, as well as mapping and monitoring the species to be found on Jinyun Mountain. This event marked the start of the journey of the nature notes program getting into the school campus and becoming a part of the school curriculum.

The Enthusiastic Participation of Students and Significant Achievements Since the launch of the nationwide Nature Notes Contest, the school has made the nature notes program a required component of the comprehensive practical activity-based curriculum. Students in both the first and second years of junior high school and senior high school are required to take the curriculum. At the same time, some of the school's elective courses also integrate the nature notes program. As a result, the majority of the school's students have participated in and completed some nature notes projects. The nature notes program also became an agenda item for school clubs. Currently, student clubs related to nature notes include the Bird Watching Club (observing birds in all their forms), the Insects Club (observing a variety of insects), the Snake and Frog Journal Club (focusing on amphibians), the Water Love Club (related to conservation of water and love of water), and the Hokuto Club (which observes the astronomical stars). Through the nature notes program, many students have developed the habit of observing nature. Some students still often discuss the natural phenomena they observed with their teachers after they graduate from school. For example, in the summer of 2019, a student who graduated from the middle school travelled to Vietnam, where he observed a gecko breaking its own tail when being threatened on the local island of Phu Quoc. After recording his observations of the entire process, the student returned home and immediately asked questions and shared them with his teacher: Why do gecko-like animals self-mutilate? What are the implications of this behavior for their own evolution and adaptation to their environment? To which species does this gecko belong? He consulted many materials, including foreign language materials, analyzed the distribution of gecko species on Fuguo Island, and compared them with the local geckos in Chongqing. As a traveler, he imagined that the gecko might migrate to here due to the human impact. Based on his observation records, he finished a Nature Note project entitled *traveling gecko and jumping tails*. The phrase of *traveling gecko* indicates that this gecko may have moved here unintentionally, while *jumping jail* illustrates the gecko's high adaptability to its environment.

The Program Contents

The nature notes program includes curriculum plans, class schedules, and teachers. The details are as follows:

Curriculum Plans The nature notes program promotes nature notes in three different ways. First, the school offers a research-based study course named Nature Notes. This is a compulsory course for the first and second-year middle school students (Grades 7 and 8) and the first-year high school students (Grade 10) (the school houses both middle and high school students). These students need to take the course for one hour each week. It is an elective for third-year middle school students and second-year high school students. Secondly, some of the elective courses build nature notes into content. For example, in one elective course that Ms. L oversees, one lesson happens to take place at the time of the graduation ceremony. Ms. L therefore organized students to conduct campus observation under the theme of Wonderful Campus Night. Through such a course task, students experience the diversity of life on campus and exercise different aspects of observation. Thirdly, nature notes are integrated into the school subject curriculum. The integration of nature notes with the disciplinary curriculum bridges the gap between disciplines and also overcomes the barrier between inside and outside the classroom. Among the disciplines that have better fused the concept of nature notes are Biology, Geography, Chemistry, Languages, and the Arts. Language course emphasizes communication through text. Biology emphasizes biodiversity. Geography emphasizes the interpretation of natural phenomena, astronomy and geology. The Arts subject emphasizes the understanding and usage of color. Through the integration of nature notes with subject contents, subject teachers create opportunities for students to learn content from multiple perspectives and enrich their relevant learning. Students can also experience life first-hand and recognize the relationship between human beings and nature. For example, when the language art teacher W taught the "Wisteria Falls" unit, she did not just follow the traditional teaching design that focused on the text's grammar, rhetoric, and composition background. Instead, she guided students to observe the flowers from a distance and then take a close look at the flowers for its fine characteristics. After students had the first-hand experience with the flowers, the teacher led the students to analyze the text and feel the meaning and emotion the author expressed in the text. In other examples, Biology teacher L paid more attention to encouraging students to observe and think about living things instead of only delivering textbook content. Similarly, Geography teacher X used the student nature notes project as teaching content. She guided the class to understand the natural phenomena such as habits and the distribution of living things recorded in some award-winning projects in terms of geology, climate, and social context. Art teacher W instructed students on how to match colors in the creation of nature notes works and observe things carefully and express them with a brush. Music teacher Z instructed students on how to record the sounds of nature. In particular, the teacher instructed a senior student to record the sound of a special toad and study its spectrum. With the help of university instruments, the student found that it was possible to differentiate males and females by acoustic spectroscopy. In short, teachers from different disciplines have integrated the nature notes program into their respective subject curricula.

Class Schedule In terms of the nature notes program timetable, students in the first two years of middle school (Grades 7 and 8) and the first two years of high school (Grades 10 and 11) take the course (comprehensive practical activity-based course) for one hour each week. In this course, the teacher will introduce the concept of practical activities, find out why they are important, the procedure and requirements of carrying out practical activities, the methods of project implantation, and how the school will scaffold their projects through organizing trips. The teacher will also recommend some project topics. In a course with nature notes as the theme, the teacher will discuss how to choose the topic for nature notes, observe the selected topic from different angles, organize one's observation records, and present such observations on A4 paper. The teacher will also promote the Nature Notes contest and encourage students to participate. The teacher often showcases some award-winning projects from past contests.

In addition to the comprehensive practical activity-based course in school time, students are required to complete an elective course during the winter and summer holidays. This elective course requires students to work on a Nature Notes project. It is designed to encourage students to use the holiday time to get close to nature. As one of the activities, some teachers will organize field trips to nature observation sites.

Teacher Resource The nature notes program is mainly implemented in the comprehensive practical-activity-based course. This course is taught by different subject teachers, including Biology, Chemistry, and Geography. Students can choose the session based on their interests. The teacher acts as an organizer, guider, and participant. They will keep regular contact with students to provide guidance. They initiate collaboration with experts from the local university so that students will get extra guidance from the experts and conduct lab activities in university labs. For example, ZW and her teammates participated in the Saddle Brook water project which lasted for three years. She often went to the laboratory at the College of Life Sciences, Southwest University to conduct water experiments during the years 2016–2018.

Nature Notes in Canada–China Reciprocal Learning Program

In 2014, a Windsor–Chongqing sister school network was built among the Canadian and Chinese schools to promote exchanging ideas between teachers and students from both countries, especially in science and technology and to enhance students' scientific inquiry skills and awareness of environmental protection (Connelly & Xu, 2019). The science team of this program first organized schoolteachers and students from Canada and China to share their classroom teaching and learning practices around an inquiry on selected science topics. Later, at the end of 2015, the sister schools identified the nature notes program implemented in the Chinese school as a useful framework for more focused collaboration and exchange. Students from both sister schools were encouraged to observe and record nature at the time of chosen solar terms. Students chose topics of interest according to the natural environment of their own locations, observed and recorded nature from at least three aspects – plant growth, animal activities, and the relationship between humans and nature – and recorded the results of their observations in photos, drawings and text. Once the program around a solar term was done, students and teachers from the sister schools shared their works through Skype video conference, including a description of the exploration process, presentation of their nature notes projects, and exchange of questions of interest. Participants of the exchange events included the science teachers and students from both sister schools, the science team researchers (Canadian and Chinese faculty members), and graduate assistants from both countries.

The research teams help to facilitate mutual communications by utilizing the Blackboard online platform and regular Skype meetings. The Blackboard platform allowed teachers and students from two sides to upload pictures and resources and exchange learnings. Students and teachers from both sides also attended regular Skype meetings where they could exchange work that had been done and the findings of each project.

Project Sharing between Two Sister Schools

During the first year of the program, two video conferences were held, and two exploratory activities were shared between the partner schools. This was a preparatory year which allowed the teachers and students from both sides to develop an initial working relationship. Between September 2016 and May 2018, the learning space was successfully built and regular communication pathways were set up for both sides. Five collaborative activities took place, and six video conferences were set for students to exchange their research progress and demonstrate their completed work (Fig. 3.7). The teachers also took these chances to build rapport, enhance further understanding of each other's culture, and exchange thoughts and ideas to make the program a better learning experience.

Nature Notes Projects from Chinese Participants

This section will illustrate the implementation of the nature notes program in Chinese sister school using student ZW as a typical case.

ZW, who graduated from the Chinese sister school in 2019, actively participated in the Nature Notes Contest during her high school years. Her work, New Discovery on the Black-eyed Toad, won the Grand Prize in the 4th Chongqing Nature Notes Contest. Her work, Transformation Axis from Larva, Chrysalis, to Butterfly, won the first prize in the 4th Chongqing Nature Notes Contest. ZW's work on black-eyed toad marked the first time of record of this species in Chongqing. She published a paper to report her discovery. ZW was selected into the Chongqing Youth Innovation Talent Development Program. Despite her busy class schedule as a high school student, she often went to the university labs to research

Fig. 3.7 Virtual exchange conferences between two sister schools (Canadian site)

the pollution and management of rivers around her school. With guidance from university professors, she published her research in a paper.

Proposing a Topic

Identifying a research topic is the first step in any nature notes project. Teachers will discuss with students the general principles of choosing a topic for their research. Students are encouraged to determine their interested subject from their experiences in daily lives and community-based activities. Sometime, students may take a topic suggested by the teacher. When helping students to choose a topic, teachers follow the following principles.

Respect for Students' Personal Interests Interest is an important motivation that drives people to understand things and explore the truth. After the foundation of Comenius and Rousseau, and the initial development by Herbart (1806) and Dewey (1913), interest in learning has become a core concern of primary and secondary education in theory and practice (Dewey & Ren, 2005; Herbart & Li, 2010). In China, Confucius, the great educator from more than two thousand years ago, is a model of *learning without boredom and teaching without tiredness.* Confucius said that the desire to learn set him apart from others. In identifying a research topic, teachers in SWUAHS respected students' individual research interests and patiently guided students to make an appropriate decision. For example, student ZW choose to study the black-eyed toad because she was inspired by its sound during an after-dinner walk. Since that time, she had become involved with the black-eyed toad and begun her own observation journey.

Teachers Guide at the Right Time In-school activities are an important part of the nature notes program and are a joint effort between teachers and students. During the topic selection phase, teachers allow students to choose a topic through purposeful and planned activities based on individual student interests and learning characteristics. The teacher will generally not interfere with the students so that they can choose what they want to observe or study. However, students will be guided to analyze the research value and feasibility of a topic and finalize the selection. For example, LJ, a biology teacher in SWUAHS, led students to Jinyun Mountain for a field trip with the intention of investigating plants, aiming to let students discover topics of interest. HYF, one of his students, was interested in the ferns he saw on the trip. He asked the teacher many questions about ferns such as "Are all fern leaves narrow leaves? "and "Why are the sporangia of ferns so neatly arranged?" The teacher L then encouraged him to take ferns as a research object. From then on, he started his research journey on nature notes about ferns. Another student, ZW, participated in a research field trip on water quality in Saddle Creek and noticed the pollution caused by the surrounding industry. She discussed this finding with her teacher, and both the student and teacher believed this could be a valuable topic to explore, considering its connection to the wider issue of environmental education. With the teacher's encouragement, ZW chose Saddle Brook water quality as her study topic.

Since many new students are unfamiliar with the nature notes program, teachers often showcase some of the best nature notes projects from previous years to help these students understand the program. In particular, it should be made clear that the nature notes program requires students to go outside of the classroom to observe, investigate, experience, and feel nature. Generally speaking, students should choose a topic that is relevant to science education and environmental education. When they have an initial idea, students will discuss their project topic with their teachers, who will then provide suggestions about what to do next. If necessary, the teacher may refer students to specialized subject teachers who can offer specific guidance.

Participation in Club Activities SWUAHS has many student clubs and teams, such as literature club, chess club, photography club, art club, math club, physics club, chemistry club, basketball team, and soccer team, among many others. Student HYF chose the red-headed and long-tailed mountain sparrow as the focus of the nature notes research after learning about birds and animals through participation in the bird watching club. Student ZW participated in the animal club. She learned about the red-spotted toad on one field trip organized by the club and later chose it as her study subject.

Observation During Out of School Parent–Child Activities Phenomena observed in daily life experience are another source for students to draw on when selecting their study topics. For example, ZW, the grand prize winner in the fourth Chongqing Nature Notes Contest, found a toad when she was walking with her mother and brother in the neighborhood park during the holidays. To her surprise, the toad did not jump when she touched it. In Chongqing, there is a local statement that *toad, touch it, and move it*. She felt very strange, so she took a picture of it, sent it to the instructor, and asked the teacher why the toad did not jump. When the teacher saw the photo, he was pleased because, a month earlier, another student had reported the existence of such toads in the neighborhood. Under the teacher's guidance, ZW found out that this toad was a black-eyed toad, which lives in South China. Later, ZW and his brother continued to observe the toad and explored how toads from South China were beginning to appear in Chongqing. After talking to many people who lived in the neighborhood, she learned that South China pond scenery had been introduced into the neighborhood for ornamental purposes and the tadpoles of black-eyed toad were transported here with fish fry. ZW recorded her observation and her findings about this toad in her nature notes project (Zhao et al., 2018).

Project Study Plan

Once the project topic is selected, students need to develop a plan to carry out the project, including the purpose and procedure of the study. Teacher CHB from Shanghai Hongqiao High School has developed a proven research methodology for nature notes projects: Three Levels and Nine Methods (Xu, 2018). The 3 × 9 matrix includes observation, recording, and reading (level one); presentation, communication, and cooperation (level two); and practice, experience, and research (level three). This methodology emphasizes implementing interdisciplinary learning content with real-life tasks and encourages students to use this research method to understand, analyze, and solve problems.

As a high school student, ZW could not observe the black-eyed toads on a daily basis. She developed a practical plan that took into full consideration the time, place, ways and means of observation, and the content of the observation.

Establishing the Study Purpose After determining the topic, students should search through the literature and clarify their research purpose based on the existing research. For example, when student HYF studied red-spotted toads, he learned through a literature search that common toads often have light green ridges in the middle of their back and dark brown stripes on the maxillary margin and limbs. The ventral surface of both sexes is creamy white, and generally without spots, although a few have small black scattered spots. Males are mostly olive-yellow on the back, with irregular spots and red spots on warts. Females are light green on the back, with saucy spots and red spots on warts, but red spots on the back are rare. HYF set the purpose of his study as (1) to let more people know about the red-spotted toad and (2) to let more people protect the red-spotted toad. HYF set up a similar purpose for his study of red-headed and long-tailed sparrows. He learned from the literature that the most famous mountain sparrow, the tufted sparrow, is 17 cm (6.5 inches) long, light blue-gray on the back, and pinkish-brown on both sides. The sparrow is usually white-faced and black-headed. It has yellow underparts in the western subspecies and whitish or yellowish underparts in the eastern subspecies, although both have black midlines. The red-headed and long-tailed sparrows are comparatively rare. He set the purpose of his study as (1) to let more people know about the red-headed and long-tailed sparrow and (2) to let more people protect the red-headed long-tailed sparrow. When ZW saw the water pollution of Saddle Creek, she decided to study water quality. The purpose of her study included (1) analyzing the water pollution (2) raising awareness of the issue among the public, and (3) pushing the authorities to take measures to protect the water body.

Deciding Study Content and Procedure Based on the proposed study purpose, students need to refine what they need to do to achieve this purpose. For example, after student HYF decided to study red-spotted toads, he developed the following study steps: (1) go online to find out what red-spotted toads are and what their habits are, (2) arrange some time every day during the summer and winter school breaks to go to Jinyun Mountain to carefully observe the red-spotted toads for a deeper understanding of their habits and reproduction, (3) present study findings, conclusions, and recommendations. After student YYH chose curly-leafed strawberries as his study topic, he made the following research steps after a literature review and teacher consultation: (1) observe the curly-leafed strawberries at home every day, (2) measure the growth of strawberries and temperature, and (3) interpret the growth pattern of strawberries.

Completion of Nature Notes Work A complete nature notes work consists of seven elements: time, place, weather, recorder, title, text, and drawing (graph or photo). Of these seven elements, four elements (time, place, weather and recorder) are the core elements of scientific documentation, which cannot be omitted. The text and drawing are the main body of the nature notes work and should be described as objectively as possible (Wu, 2018).

Time: In a broad sense, different years and seasons have different phenology, and recording the time helps to carry out comparative studies.

Location: Species vary from region to region, and the same species has different phenotypes or behaviours in different regions. The information about the location helps the researchers to detect patterns relevant to geographical locations and also makes it possible for further study on the same phenomena at the same location.

Weather: Natural phenomena vary significantly in different weather conditions. The information about weather is a factor to consider when we study nature.

Recorder: Nature observation is subject to the differences of individual recorders. Observers bring their own intention, interest, focus, and perspective into the observation behavior and results. The information about the recorder helps the readers to understand the nature notes work. It also makes it possible for communication to be carried out between researchers.

Title: The title reflects the content of the nature notes work and it should be vivid and attractive to the readers. For example: *Cat Walk Cat Style* by student ZDT and *Summer Singer* by LH are good examples of attracting tiles.

Text: Text is the main body of any nature notes work. The observer can obtain much information by observing nature from multiple perspectives through multiple senses. Text-based descriptions of the natural phenomena should be objective and realistic, rather than fabricated. Students are allowed to use metaphors in their text considering younger student participants may lack the specialized vocabulary. Participants from lower grades can use phonetics to substitute the words they do not know how to write.

Drawing: Drawing supplements text descriptions as a visual representation of natural phenomena. It should be as realistic and objective as possible, rather than exaggerated or overstated. All drawing must be done to scale. Besides rulers, students are encouraged to use their fingers, pens, footsteps, or nearby trees and buildings to estimate the size of the observed subject. A close-up view of a particular part can be used to show details. Drawing should not include cartoonish figures and other non-objective depictions, nor should they include the drawing of eyes on fruits or lace on leaves. There should not be any artistic rendering effects such as background color or underline.

The nature notes works require students to present their observations and perceptions on a single sheet of A4 paper. This is quite challenging since students must determine what information to include in such a limited space. For example, student ZW collected many materials about Jinyun Mountain toads and she needed to select the representative and typical contents for his final presentation of her work. In the end, she decided to focus on the group sound of the toads as the main focus of her work presentation. Then she designed the presentation layout, font style and size, and graphic styles to nicely fit the chosen content into the display space one sheet of A4 paper offers.

Showcase Achievement Students have actively participated in the past Nature Notes contests, with many excellent works standing out. Through these works, the participants' meticulous observations, exploration, thinking, and meaningful expressions can be seen. Take the first prize winner in the sixth Chongqing Nature Notes Contest, *Love and Hate Water Gourd*, as an example (Fig. 3.8). The author of this work introduced the structural characteristics of the water gourd with beautiful paintings and logos, giving people the most intuitive understanding. The first prize winner of the 5th Chongqing Nature Notes Competition, *Cat walk Cat Style*, showed the stereotypical behaviour of golden cats in zoos, in contrast to leopard cats (Fig. 3.9). Through observation, the student found that the living environment can affect the behavior of animals. Since the golden cat is a group living animal, keeping it alone in the zoo caused it to feel lonely and develop stereotypical behaviors. Such observation led the author to profound reflections on how we should treat wild animals and explore the idea of building wildlife parks and national parks. Another award-winning

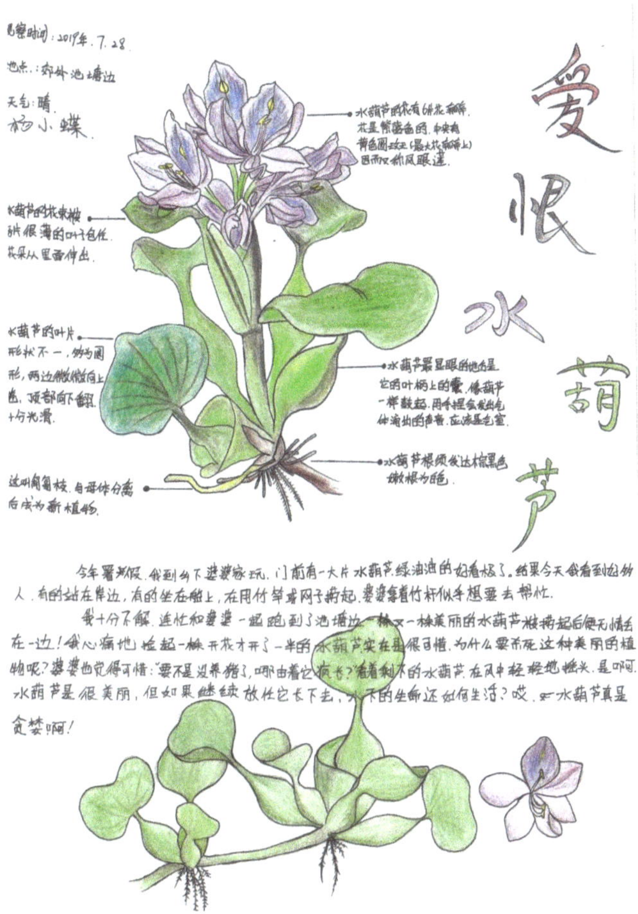

Fig. 3.8 Love and Hate Water Gourd. Note: The first prize winner of the 6th Chongqing Nature Notes Contest (author: Xiaodie Yang, Ruishan Middle School, Hechuan, Chongqing. Mentor teacher: Hui Mi). Judge Committee Comment: Nature is not only in the wilderness and faraway places but also around us and under our feet. Xiaodie's nature notes record the water hyacinth by the pond. She loves its beauty and hates its crazy growth that hinders the growth of underwater creatures. Xiaodie meticulously depicts the flowers, leaves, branches, roots, and air chambers of the water hyacinth, giving a visual sense of its characteristics

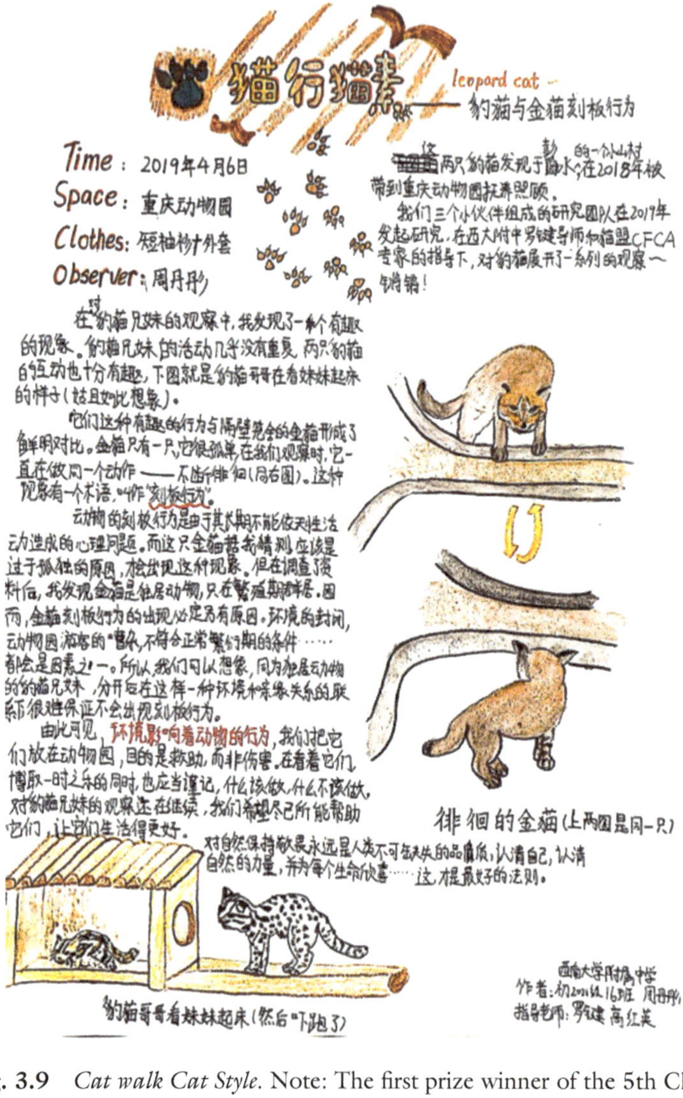

Fig. 3.9 *Cat walk Cat Style*. Note: The first prize winner of the 5th Chongqing Nature Notes Contest (Author: Dantong Zhou, Southwest University Affiliated Secondary School. Mentor teachers: Jian Luo and Hongying Gao). Judge committee comment: The young author explored the reasons behind animals' behavior differences, which reflects his inquiry intention and ability. He encouraged to include more observation records and less theoretical knowledge in future work

work, *Summer Singer*, records the students' observations of the black grasshopper for two consecutive days, with vivid and detailed drawings of the process of egg-laying (Fig. 3.10). In the second prize entry of the 5th Chongqing Nature Notes Contest, *Bayu Horned Monster*, the student introduced the discovery of the Emei Horned Toad at the Autumn Equinox in Wulong, Chongqing, a karst landscape in southern China (Fig. 3.11). With delicate brushstrokes, the students drew the geographical location of the Chongqing sinkhole, the morphology of tadpoles, female and male horned toads, and the seven distribution areas of the horned toads. The student described the male horned toad's appearance using interesting language such as "the eyes look like pupils, and the moustache on the male's lip stands out." The student also learned that the Emei horned toad was an endangered species, and her emotion for protecting the ecological environment and endangered animals was aroused.

Communication and Evaluation Communication and evaluation activities among students and between the teacher and students are essential parts of the nature notes projects. Students can learn from and help each other through evaluating each other's work for strengths and weaknesses. Teacher–student communication allows students to ask for advice from the teacher at every stage of their projects.

Nature notes works are evaluated in multiple steps. First, the teacher who supervised student projects made preliminary evaluations of student entries and selected the best entries for the Nature Notes Contest. Then the judge committee of the Nature Notes Contest graded the entries. Each year, the contest set up one grand prize, the 1st, 2nd, 3rd prizes, and merit awards. The purpose of these prizes and awards was to recognize students' outstanding works and encourage students to get in touch with nature and improve their observation and research skills. Student ZW said after receiving one award, "I received a title of an excellent student with a certificate of completion. My teacher gave me high praise for my performance, and I am thrilled to be part of the nature notes program."

Fig. 3.10 Summer Singer. Note: The first prize winner of the 5th Chongqing Nature Notes Contest (Author: Hang Li, Yinxiang Experimental Middle School. Mentor teacher: Huanxia Chen). Judge committee comment: The young author observed the egg-laying behaviour of cicadas very carefully, with rich textual records and vivid details. The beautiful and accurate hand drawing complements the textual content. The typographic organization of the text can be further optimized

Implementation of the Nature Notes Program in the Canadian School

The Nature Notes Reciprocal Learning Programs

The Nature Notes Reciprocal Learning Program was designed to foster ideas exchange, share teaching and learning experiences, and explore the cultural impacts between the Southwest University Affiliated Secondary School in Chongqing, China and the Prince Edward Public School in Windsor, Canada. Based on the expectation of developing reciprocal learning relationships, these two schools agreed to collaboratively conduct the nature notes program in 2015. The two schools conducted the program around the same solar term but did it differently according to each school's situations. The teachers from both sides selected an agreed date

Fig. 3.11 Bayu Horn Monster. Note: The second prize winner at the Fifth Chongqing Nature Notes Contest (Author: Yaru Lei, Chongqing People's Primary School. Mentor teachers: Jian Luo and Hongying Gao). Judge committee comment: This is a rich and rigorous nature notes work. In the process of recording the Emei mustached toad, the author not only explained the habitat where it was found but also compared the differences between male and female individuals and the characteristics of juveniles. In addition to these personal experiences, the author also researched the distribution of the Emei mustard toad and the history of the

to start each project cycle. The dates typically fall at the beginning of each selected solar term. Students and teacher from both sides exchange their progress or findings periodically.

This partnership program has run for three school years, from September 2015 to May 2018. Teachers from both sides chose a solar term to conduct the project. Two to three project cycles happened each school year. The choices of observation time for each project cycle refer to the Chinese solar terms, which specify the 24 points where the weather noticeably changes around the year. The Canadian students were taken out to observe nature, record their observations, and collect what interested them. Then, each group of 2–3 students chose a topic of interest to work on. The choice of topics was fully driven by the students' personal interests. Teachers from both sides designed the nature notes projects according to their curriculum and based on their class schedules and situations. Each project cycle refers to one solar term, and includes three stages: preparation; observation and data recording; and presentation. The content, time and purpose of each stage are shown in Table 3.2.

The Canadian teachers decided that integrating the nature notes program into their curriculum and making it an inquiry-based learning experience was the most appropriate way to implement it based on their teaching schedules. In other words, they combined two instructional objectives into one project, and students had to meet a success criterion that covered both aspects. To increase the effectiveness of the program implementation, the instructor designed a resource package (Appendix 1) for students to organize their ideas and perspectives and prioritize their work. Practice sheets were provided to students as a guide. The work packet was given to each student, and they could take it home. Teachers provided guidance to student groups for their interested study topics and research plans. The inquiry-based learning packet covers the following components: selecting a topic and planning for investigation, retrieving information, evaluating the sources, processing information, listing a bibliography, and assessment rules (for self-reflection). The teacher chose a

species when it was first discovered. The work is great! We often say protecting nature. The prerequisite of nature protection is to understand nature. We need to know what we want to protect and how we can protect it. Without personal experiences with nature, it is hard to build a connection between humans and nature and it will be hard to take measures to protect it. Consistently writing nature notes can help us get close to nature, learn about nature, and gradually build up a sense of concern for environmental issues, just as this young author expressed in her work!

Table 3.2 Steps of project cycle

Stage	*Content*	*Time*	*Purpose*
Preparation	Study the sun clock, and learn about the historical development in the 24 solar terms and the 12 constellations. Find out the similarities and differences between the two	Five days before each project cycle.	Teachers and students understand the relationship between the 24 solar terms and Chinese farming culture, and its significant impacts on the Chinese agricultural activities.
Observation & Data Recording	Through daily observation and information checking, the Chinese and Canadian students individually record the local weather, temperature, humidity, dressing index, sunshine duration, and air quality. Both sides select a similar plant and observe its root, leaf, flower and fruit. Use daily log to record a story of a favorite animal or seasonal vegetables and fruits that the students like.	At the beginning of each project cycle.	During the sharing sessions, the two sides compare the similarities and differences of the plans and animals that they observed. As a return, they build a deeper understanding of each other's environment and culture.
Presentation	Exchange study findings conducted by sister schools. Use charts, photos, and text to present student observation and research on local weather, plants, animals, and human activities.	At the end of each project cycle.	Gain a deeper understanding of each other in terms of the school context, and the natural and cultural environment. Build and maintain a friendly relationship with teachers and students from the partner school.

day around one solar term to be the start day of the project. On this day, the teacher brought the students to the campus to observe (Fig. 3.12).

During the nature observation, students were equipped with the necessary tools (such as gloves and containers) to collect materials for future use, and they also took the opportunity to experience nature and record the necessary information. The local weather, temperature, humidity, clothing index, sunshine hours, and air quality were then filled in on a nature notes observation form. Next, the collected information was organized, and students chose a topic based on their interests. In the third

Fig. 3.12 Canadian Students Observe on School Campus

step, each student chose a broad area of investigation and then proposed a question they would like to research (fill in on the project plan sheet). In step four, students chose a way to present their learning and signed an agreement with the teacher to ensure that the project was completed within a set time frame. In step five, the teacher asked two questions, and the students chose to answer one of them as a final reflection.

The inquiry package was designed in a well-organized format and provided a systematic way to help students complete their project. It helped students assure their commitment to completing the project by a deadline. What impressed the students most was that they had the freedom to choose their topic and demonstrate their work in a way they wanted to. In other words, the project was mostly driven by the student's personal interest, and students enjoyed this flexibility. The students were extremely satisfied because they had never before experienced such complete freedom in other assignments. This feeling was essential to the students and consequently kept them highly motivated to participate in and complete the project in a timely manner. The creativity of their products, the confident attitudes shown in the video conferences, and the satisfaction demonstrated in the interviews together lead to a conclusion that the Nature Notes Reciprocal Learning Program was successfully implemented by the teachers in the Canadian school.

Appendix 1: 24 Nature Notes Project Inquiry Working Package from the Canadian School

Nature Notes Curriculum Expectations

Oral Communication

Overall:

1. use speaking skills and strategies appropriately to communicate with different audiences for a variety of purposes;
2. reflect on and identify their strengths as listeners and speakers, areas for improvement, and the strategies they found most helpful in oral communication situations.

Specific:

2.1 identify a range of purposes for speaking and explain how the purpose and intended audience might influence the choice of speaking strategies

2.2 demonstrate an understanding of appropriate speaking behaviour in most situations, adapting contributions and responses to suit the purpose and audience

2.3 communicate orally in a clear, coherent manner, using a structure and style appropriate to both the topic and the intended audience

2.4 use appropriate words, phrases, and terminology from the full range of their vocabulary, including inclusive and non-discriminatory language, and a range of stylistic devices, to communicate their meaning accurately and engage the interest of their intended audience

2.5 identify a range of vocal effects, including tone, pace, pitch, volume, and a variety of sound effects, and use them appropriately and with sensitivity towards cultural differences to communicate their meaning

2.6 identify a variety of non-verbal cues, including facial expression, gestures, and eye contact, and use them in oral communications, appropriately and with sensitivity towards cultural differences, to help convey their meaning

Writing

Overall:

1. generate, gather, and organize ideas and information to write for an intended purpose and audience;
2. draft and revise their writing, using a variety of informational, literary, and graphic forms and stylistic elements appropriate for the purpose and audience;
3. use editing, proofreading, and publishing skills and strategies, and knowledge of language conventions, to correct errors, refine expression, and present their work effectively;
4. reflect on and identify their strengths as writers, areas for improvement, and the strategies they found most helpful at different stages in the writing process.

Specific:

1.1 identify the topic, purpose, and audience for more complex writing forms

1.2 generate ideas about more challenging topics and identify those most appropriate for the purpose

1.3 gather information to support ideas for writing, using a variety of strategies and a wide range of print and electronic resources

1.4 sort and classify ideas and information for their writing in a variety of ways that allow them to manipulate information and see different combinations and relationships in their data

1.5 identify and order main ideas and supporting details and group them into units that could be used to develop a multi-paragraph piece of writing, using a variety of strategies and organizational patterns

1.6 determine whether the ideas and information they have gathered are relevant, appropriate, and sufficiently specific for the purpose, and do more research if necessary

2.1 write complex texts of different lengths using a wide range of forms

Reading

1.1 read a wide variety of increasingly complex or difficult texts from diverse cultures, including literary texts, graphic texts and informational texts

1.2 identify a variety of purposes for reading and choose reading materials appropriate for those purposes

1.3 demonstrate understanding of increasingly complex texts by summarizing important ideas and citing a variety of details that support the main idea

1.4 identify the point of view presented in texts, including increasingly complex or difficult texts; give evidence of any biases they may contain; and suggest other possible perspectives

Nature Notes: The Awakening of Insects

The "Awakening of Insects" is one of the cycles in the 24 solar term. At that time, there is plenty of rainfall, indicating spring is coming. Ancient Chinese people believed that the thunder wakes up all insects, and people should begin their agriculture affairs.

During the period of "Awakening of Insects", the temperature warms up, and in spring there is usually plenty of rainfall. In addition to the northeast, while everywhere is still covered with snow, the average temperature in North China for the day is 3-6 °C, but the average temperature in most parts of China has risen to more than O °C. For example, Jiangnan has the temperature of more than 8 °C while the Southwest and South China has reached 10-15 °C.

In South China Spring fills the air with warmth while north China is still freezing cold. Therefore, China attaches great importance to working people since in ancient times the Awakening-of-Insects cycle creates a spring start date. It is the beginning of agriculture, and farmers begin to farm their land and produce crops.

Information from article on Chinese Customs http://cn.hu jiang.com

Your Assignment for the "Awakening of Insects"

1. Go outside, and make your observations (fill out Nature Notes Observation Sheet).

2. Choose a topic of interest on either insects or animals.

3. Choose your broad area of inquiry, and then create an inquiry question that you will research (fill out your project planner).

4. Choose a method in which you want to demonstrate your learning (Fill out your Nature Notes Contract).

5. Choose one of the following questions, answer it once you have completed your research, and either attach it to your good copy, or incorporate your answer into your project itself.

    1) What lesson did you learn about life through understanding your chosen insect or animal?

    2) What do you see differently now that you know about your insect or animal?

Project Planner

Selecting a Topic and Planning an Inquiry

Outcome 2.1.1

What is my broad area of inquiry?

Narrowing the focus…

Some possible inquiry questions…

Where can I find reliable information sources?	Who will be the audience and what format will I make my presentation?	How will I be evaluated on this inquiry project?

What is my plan and schedule? Include checkpoints.

Guided Practice

Selecting a Topic and Planning an Inquiry

Outcome 2.1.1

How do I select a topic and plan my inquiry?

Brainstorm ideas and ask questions that interest you. For example, if you want to know more about "child soldiers," then you need to come up with a number of questions that interest you. This will help you narrow the focus to something that can be researched. Remember- you are trying to answer a question that has not been asked before- not just looking for someone else's answers. As you search for sources that relate to your question, you could find that your inquiry question changes or needs to be refined more.

TIPS: Web Searches

GOOGLE is a search engine, not a website or a source that can be cited in your research. It is a good starting place to get ideas, but do not rely totally on this for your research.

Wikipedia may be tempting to use too but is not always reliable as a source, and it should be viewed only as a starting point – a place to find idea and additional sources at the end of each article.

Guided Practice:

Enter "child soldiers" into an online search engine. Notice how many possible links there are for this topic- obviously, we need to narrow the topic!

Broad Topic:	Child soldiers
Narrower Topics:	Recruitment of child soldiers, gender-related experiences, organizations and efforts to reintegrate children into society, international laws related to child soldiers
Possible Question:	How do former child soldiers move back into society?
Possible Sources:	Websites, encyclopedias, journals, and other sources that can provide reliable information. Use a variety of formats
Audience:	Class/teacher/community/other
Format of Presentation:	Digital presentation, mini-documentary, photoessay, research paper, oral presentation, talk show simulation
Evaluation Criteria:	Teacher and/or student-generated criteria to evaluate product AND process (includes "learning to learn skills")

Guided Practice

Retrieving Information

Outcome 2.1.2

How do I go about retrieving information for my inquiry?

Searching for information can be a tough job for even the most experienced researcher. Stay organized and keep a record of your searches. You will likely need to find these sites again. Start by planning out your search. You might assume that the World Wide Web is the best place to begin but there are lots of other options too. Online searches can be time-consuming and frustrating. Try out encyclopedias, texts, videos, periodicals (magazines), and databases such as EBSCO; they are right within reach at school, home, or your local library. There are often community sources as well such as government records or materials produced by community organizations. Don't forget to ask for help!

TIPS: Primary and Secondary Sources

Primary sources are first-hand materials such as a novel written by an author, letter, diary or journal entry, autobiography, speech, personal interview, first-hand account of an event, photograph, painting, or other original work. **Secondary** sources include all second-hand accounts of primary sources or materials that have been interpreted by others-movie and book reviews, text books, translations, encyclopedia articles, historical accounts (written by someone who was not there at the time of the event), or recreated artifacts or replicas. Sometimes it is difficult to tell if a source is primary or secondary (and, may in face be a bit of both). In the case of web searches, articles on a specific topic with a seated author are generally primary sources, but these would be considered secondary if the article interprets work that has already been published.

Guided Practice:

1. Make a checklist of all the possible places where you might find information.
2. Keep detailed records of the sources you find that you intend to use. If a source is not a good match, discard the record to avoid confusion.
3. Look closely at the URL addresses of any websites that you may use - URLs hold clues to reliable sites or ones that may be biased. Enter "child soldiers" into a search engine such as GOOGLE and note the domain tag on the URLS (this is the 3-lecter clue to the origin). For example, "edu" refers to an educational organization or institution; "org" refers to a (usually) non-profit or governmental organization; "gov" refers to________________: and "com" means the site is________________.
4. Scroll through the first 20-30 hits for "child soldiers" and see how many fit the four categories above: edu ____; gov ____; org ____; com ____.

Guided Practice

Evaluating Sources

Outcome 2.1.3

How do I know if a source is a good one?

While you may think that you'll never find enough material to complete your project, more often the complete opposite is the case. Finding sources is one thing–finding good sources is a whole other thing. Just as important as knowing a bit about the author and the intended audience of the information is being certain that the information is relevant to your work.

TIPS: Plagiarism

Avoiding plagiarism can be tricky when you are selecting information. If you are using data, findings, arguments, or any other work of others, you must give credit to that source. For example, if you are using statistics about the number of child soldiers worldwide, or research results about the psychological impact of war on children, you must cite the source. Common knowledge need not be cited as it is generally shared by all readers (e.g., recruiting child soldiers is a violation of their human rights). If you are not sure, check with a teacher or librarian, or refer to a writing handbook for more guidelines.

Guided Practice:

1. Ask yourself if the material is closely related to your inquiry. For example, if your topic is about how child soldiers are rescued and move back into society, you need not include information about the climate of their country, or the national sport.

2. Use the same guidelines for measuring reliability of the author as you as you would for selecting sources. Nationally-known organizations, educational institutions, or expert authors are the most reliable sources.

3. Is the information up-to-date, or, is there historical data that may be useful? Older materials may prove to be valuable but check to be sure that the data is still current and has not been replaced by newer information.

4. Is the material easy to access when you need it? Remember to record it!

5. Is there an obvious bias or does the article present a balanced view?

6. Is there enough material to help out your inquiry? Or, is there so much that you need to be selective in matching it to your work? Select three sources of information on child soldiers and try to answer the guiding questions above to get a sense for how relevant or valuable the material is to you.

Guided Practice

Processing

Outcome 2.1.4

Now what? How do I pull it all together?

By now, you've gathered a number of sources of information for your inquiry. You've done some weeding, sorted through materials, and already learned quite a bit. Now, it's time to finalize your focus and select the most relevant information. You may find that you've shifted your focus a bit as you came across new information and changed paths. That's all part of the inquiry process and shows that you are constantly evaluating and re-evaluating information. At this point, you may discover that you either need to narrow your focus or broaden it somewhat to capture what it is you want to find out about this topic.

TIPS: Note-taking and Summarizing

Being able to take good (not necessarily lengthy) notes and summarizing information is a skill that will benefit you for a lifetime-but it takes practice and patience. Some people like to use a note-card system or other means that works for their particular styles. The main thing is that you stay organized and efficient.

Do...

- record the source, author, and page somewhere obvious.
- copy quotations exactly as they appear in the original.
- summarize ideas in your own words-see below.

Summarizing v. Paraphrasing

- Both mean restating original work into your own words.
- Summarizing reduces the original ideas by at least 50% whereas paraphrasing will be about the same length as the original.
- Concept maps are a form of note-taking and can be very detailed.

Guided Practice:

1. Select an appropriate source of information on a topic such as child soldiers or a variation of this topic and summarize a principal paragraph or section. Use a SQ3R strategy - survey, question, read, recite, review to help you. Ask your teacher for assistance on this strategy if you aren't sure.
2. Try to reduce the original paragraph or section by half using your own words and the key ideas of the section.
3. Share with a partner to evaluate how well you did in summarizing.

MLA Bibliography
Garfield and Harding Middle Schools

A "Works Cited" or bibliography is the list of works that you used during the research phase of your paper. There are two purposes for creating a bibliography: 1) to give proper credit to your sources and 2) to allow your reader to locate your sources. Therefore, your major objectives when creating a citation are to be accurate and complete. An on-line tool called EasyBib.com can also help achieve accuracy.

The sources on your "Works Cited" page must be listed alphabetically by the last name or first word of the entry. It is VERY IMPORTANT that you mind the arrangement and punctuation format of each entry-these are not open to interpretation! Notice that the second line (and each subsequent line) is indented five spaces.

Print Resources (Viewed in Print Form)

Book by One Author

Last Name, First Name. Title. Place of publication: Publisher, Copyright Date.

Example:

Jameson, George P. Ellis Island. New York: Icon Press, 2006.

Book by Two Authors

Last Name, First Name and First Name Last Name. Title. Place of publication: Publisher, Copyright Date.

Example:

Smith, Henry G. and Betty Harmon. Freedom Rides. Chicago: Broad Shoulder Press, 2006.

Encyclopedia (NOTE: The author is usually listed at the end of the article. If an author is not listed, begin the entry with the title of the article.)

Last Name, First Name. "Title of Article. " Name of Encyclopedia. Edition Year.

Example:

Franz, Charlene. "Goat." World Book. 2006 ed.

A Work within a Larger Work (Anthology)

Last Name, First Name. "Title of work." Title of Anthology. Editor's First Name Last Name. Place of Publication: Publisher, Copyright Date. Page numbers.

Example:

Stevens, Wallace. "Sunday Morning ." Modern American Poetry. Ed. Thomas J. Hines. New York: Holt, Rinehart and Winston, 2003. 20-22.

Newspaper Article

Last Name, First Name. "Title of Article." Newspaper Name Date: page(s).

Example:

Blake, Terry. " Attack in Bagdad: Two Marines Dead." The Plain Dealer 20 July 2006: A1.

Magazine Article

Last Name, First Name. "Title of Article." Title of Magazine Date: page numbers.

Example:

Thomasson, Ronald. "Salt Mines of Lake Erie." Cleveland Magazine 15 May 2004: 23-25.

LRC/Costello 2008

Non-Print Resources (Viewed Electronically)

Online Database-Encyclopedia

"Title of Article." Name of Encyclopedia. Year. Name of online source. Date <URL (Persistent link)>.

Example:

"Whale Oil." Encyclopedia Britannica. 2006. Encyclopedia Britannica Online. 23 August 2006 <http://search.eb.com/eb/article-9076727>.

Online Database-Articles provided through Infohio.org Databases

Last Name, First Name. "Title of Article." Original Print Source Title. Date. Online source name. Date accessed <URL Persistent link>.

Example:

Freeman, Gregory A. "Code Alpha: The President is Coming!" American History. October 2006. Academic Search Premier . 6 October 2006 <http://web.ebscohost.com/ehost/detail?vid=7&hid= l&sid=13275eeb-239a-4ec6-a998-d2dda6dc9f66%40sessionmgr101>.

Internet Website

(Note: it is sometimes very difficult to find all of the information you need to make a proper citation for a website. Try to find as much information as possible.)

Author (if given). "Title of Web Page." Title of Website. Date last updated. Name of organization that sponsors the site. Date accessed <URL link>.

Examples:

Flannery O'Connor Collection. 7 July 2006. Georgia College and State University. 31 August 2006. <http://library.gcsu.edu/~sc/foc.html>.

Walker, Gary. "The Effects of Radiation." Hiroshima Atom Bomb. 15 Mar. 2000. Los Alamos Research Facility. 14 Oct. 2008. <http://www.larf.org/hiroshima/radiationeffects.htm>.

TV / Radio Transcript

"Title of TV/ Radio episode or segment." Title of Program. Broadcast date. Transcript. Name of Database Used. Date accessed < URL link >.

Example:

"Special Edition: The War in Iraq Part II." CNN Special Report. 5 April 2003. Transcript. eLibrary. 27 March 2006. <http://elibrary.bigchalk.com>.

Video Documentary

Title of Film. Director's First Name Last Name. Copyright Year. Medium. Production Company Name. Edition Date.

Example:

Finding Harmony. Kent Williams. 2003. DVD. Zen Buddhist Lecture Series Video. 2003.

LRC/Costello 2008

Nature Notes Observations

Observation Categories	Observations or Information Check
Local Weather	
Temperature	
Humidity	
Dressing Index	
Sunshine Duration	
Air Quality	

Observations on possible topics:

Nature Notes

To demonstrate what I have learned about ______________________, I want to

_ Write a report
_ Put on a demonstration
_ Set up an experiment
_ Develop a computer presentation
_ Build a model
_ Design a mural
_ Write a song
_ Make a movie
_ Create a graphic organizer or diagram
_ Other___________

This will be a good way to demonstrate understanding of this concept because

____

To do this project, I will need help with

____

My Action Plan is____

The criteria/rubric which will be used to assess my final product is __________

____

My project will be completed by this date________________________________

Student signature: ________________________________ Date ___/___/___

Teacher signature:________________________________ Date ___/___/___

Nature Notes

Name ____

Categories & Expectations	Level 1	Level 2	Level 3	Level 4
KNOWLEDGE & UNDERSTANDING Understanding of content	The student shows limited understanding of content	The student shows some understanding of content	The student shows good understanding of content	The student shows insightful understanding of content
THINKING Use of problem solving skills & creative/critical thinking processes	Able to -initiate own learning -solve problems -use creative/ critical thinking processes to create final product -reflect on their learning with limited effectiveness	Able to -initiate own learning -solve problems -use creative/ critical thinking processes to create final product -reflect on their !earning with some effectiveness	Able to -initiate own learning -solve problems -use creative/ critical thinking processes to create final product -reflect on their learning with considerable effectiveness	Able to -initiate own learning -solve problems -use creative/ critical thinking processes to create final product -reflect on their learning with a high degree of effectiveness
COMMUNICATION Communicate ideas orally and written	Able to represent his/her thinking orally and on paper (illustrations and/or text) with limited effectiveness	Able to represent his/ her thinking orally and on paper (illustrations and/ or text) with some effectiveness	Able to represent his/her thinking orally and on paper (Illustrations and/or text) with considerable effectiveness	Able to represent his/her thinking orally and on paper (Illustrations and/or text) with a high degree of effectiveness
COMMUNICATION Knowledge Building Circle	Able to state wonderings, present questions and listen to others ideas with a limited effectiveness	Able to state wonderings, present questions and listen to others ideas with some effectiveness	Able to state wonderings, present questions and listen to others ideas with a considerable effectiveness	Able to state wonderings, present questions and listen to others ideas with a high degree of effectiveness
COMMUNICATION Use of vocabulary, terminology, organization &. conventions	Able to use conventions, vocabulary, and terminology related to topic. Information is not organized.	Able to use conventions, vocabulary, and terminology related to topic. Information has some organization.	Able to use conventions, vocabulary, and terminology related to topic. Information is clearly organized.	Able to use conventions, vocabulary, and terminology related to topic. Information is effectively organized.

Appendix 2: 24 Nature Notes Project Inquiry Working Package from the Chinese School

An Example of Nature Notes Project Design

Course (activity) Implementation Period: Sep. 2018–Oct. 2019
Course (activity) Implementation Site: Zhangfei Ancient Road (the east bank of Jialing River, Beibei District, Chongqing)
Course (activity) Name: Learning about the Ancient Road through Notes
Subjects Involved: Geography, comprehensive practical activities (research-based learning), biology, history, literature, etc.
Instructors: Lead instructor Li Jiubin, collaborating instructor Li Jiaying
School: Southwestern University Affiliated Secondary School

	Content
Instructional (activity) objectives	**1. Cultural foundation** **Culture heritage**: Understand the cultural value of the Zhang Fei Ancient Road and appreciate the calligraphic inscriptions and poetic contents along the way; appreciate and evaluate the beauty of Zhang Fei Ancient Road; communicate the beauty through Nature Notes work. **Scientific attitude**: Advocate true knowledge and rationally analyze the site of Zhang Fei Ancient Road; propose questions, think independently and dialectically with curiosity and imagination; illustrate the process of investigation through Nature Notes work. **2. Personal development** **Learning how to learn:** Through multiple site visits to Zhang Fei Ancient Road, build interest in learning, develop good study habits, and form an awareness of independent learning and career planning. **Healthy living:** Through multiple walks on Zhang Fei Ancient Road, enjoy the physical exercise; improve awareness of safety; stay positive, confident, and resilient; have a sense of teamwork. **3. Social participation** **Responsibility**: During the site visits, follow the rules, love and respect nature, and protect the ecological environment of Zhang Fei Ancient Road, Jialing River and Jinyun Mountain; have a sense of native land and understand the natural and humanistic history of Beibei district. **Practice and innovation:** Identify questions, apply what learned to solve real-life problems, disseminate ideas in the form of Nature Notes work.

(*continued*)

(continued)

Selection and elaboration of instructional (activity) themes	**1. Theory and policy** The Chinese government promotes the idea of ecological civilization, advocating the concept of green water and green mountains as the silver and gold mountain of gold and the pursuit of harmony between human beings and nature. Implementation of such idea needs education in the real context, learning by doing through project-based learning and research-based learning. **2. Curriculum standards** The new curriculum reform advocates education that makes full use of the surrounding natural and social resources, guides students to use field trips for practical activities, cultivates core literacies around the concept of life, rational thinking, scientific inquiry, and social responsibility. **3. Surrounding environment** Beibei is backed by Jinyun Mountain and surrounded by the Jialing River. It is famous for its rich cultural atmosphere, famous scenic spots, and beautiful landscape. Zhang Fei Ancient Road is the masterpiece of Beibei's natural landscape, crossing Jinyun Mountain and adjacent to Jialing River, with many ancient trees and lush greenery. It is the preferred place to carry out nature notes projects. **4. Nurturing value** The nature notes project at Zhang Fei Ancient Road, on the one hand, enhances students' understanding and knowledge of Beibei, and on the other hand, allows students to apply what they learned in classroom to the real life and actively inquiry and explore the nature.

(continued)

(continued)

Instructional (activity) procedure

Preparation

• Identify students who will participate in the project and notify parents and administrative classroom teachers for their support. Parents will be invited to participate in the Zhang Fei Ancient Road walk.
• Review students' knowledge base and find out what students already know about the natural environment of Beibei, especially Jinyun Mountain and Jialing River.
• Determine the implementation steps of the project, specifying the purpose, content, format, location, equipment and instruments, and the necessary guidance from professionals.
• Carry out an initial visit to Zhang Fei Ancient Road for site exploration and plan what to do for the project.

Implementation

The implementation stage has two major parts: Initial acquaintance with Zhang Fei Ancient Road (part I) and re-exploration of Zhang Fei Ancient Road (part II). The specific content of each part is illustrated in the following table. Given that many students have little prior knowledge, part I focuses on students' overall understanding of the geography of the site. Through the first site walk, students will take notes (photos and written notes) of the nature beauty and record any curiosity or questions they may have. This provides a basis for part II. Re-exploration of Zhang Fei Ancient Road aims at students' deeper understanding of the site. On the basis of sharing and discussing their learning from the initial visit, students will take more site visits to take further notes for enrichment and refinery of their initial learning.

	Topics	Mission format	Class location	Lesson hours
Part I	Introduction of Jinyun Mountain and Jialing River	Group report	Classroom	2
	The history of Zhang Fei Ancient Road	Study list	Classroom	
	Initial acquaintance with the ancient road for its attraction	Photo taking	Zhang Fei Ancient Road	4
Part II	Photo sharing	Group report	Classroom	2
	Questions and discussion	Cooperative learning	Classroom	
	Re-exploration of the ancient road for its meaning	Research of study list	Zhang Fei Ancient Road	4

Reflection and dissemination

• At the end of part I, students will write a report for discussion in part II. At the end of part II, students will write a final project report.
• Collect the activity reports, pictures, videos and other related materials and organize them into a book as a resource for future school-based curriculum.
• Disseminate students' nature notes works through school bulletin boards, LED display screens, and school website for the benefit of school community. Student works will also be shared through local community for public education.
• Refer some of the best student works to the nature notes contest

(*continued*)

(continued)

Learning tasks, assignment and assessment

Learning tasks and assignment

Assignments are given at each step of the course so that the learning outcomes can be tested and recorded promptly.

	Topics	Assignments
Part I	Introduction of Jinyun Mountain and Jialing River	• Through internet and library search, get to know the climate of Beibei district, the vegetation and biological condition, geological features, and soil condition of Jinyun Mountain, and the hydrological condition of Jialing River • Group presentation of findings to the class
	The history of Zhang Fei Ancient Road	• Learn the history of the site through teacher lectures and seminars. • Develop a study list
	Initial acquaintance with the ancient road for its attraction	• Take notes and photos about the spots that students are interested in for an in-depth exploration • Draw a map of the ancient road and mark the chosen spots on the map
Part II	Photo sharing	• Share photos individually or in small groups to the class, paying attention to the content, stories, and insights behind photos • Revise the initial nature notes
	Questions and discussion	• Discuss in groups the questions students may have from the initial site walk and record the questions that needs further exploration during the follow-up walks • Revision of nature notes
	Re-exploration of the ancient road for its meaning	• Complete the topics on the study list and revisit any unanswered questions • Complete the final draft of the nature notes within two weeks of the last walk.

Student assessment

Student assessment takes into consideration the learning process, learning performance, and learning outcomes. Instead of only looking at the final products, the teacher will pay attention to students' performance in the learning process for a deeper understanding of students' motivation and their operational, analytical, and decision-making skills. Course grade is determined by the following components and weighs: Attendance (15%), class participation (30%); the road map of Zhang Fei Ancient Road (15%), Nature Notes works (25%), and contest award (15%). Where applicable, self assessment, peer assessment, and teacher assessment are cross used.

References

Connelly, F. M., & Xu, S. J. (2019). Reciprocal learning in the partnership project: From knowing to doing in comparative research models. *Teachers and Teaching, 25*(6), 627–646. https://doi.org/10.1080/13540602.2019.1601077

Dai, Z., & Jin, Z. (2007). Geographical issues in a local proverb. *Secondary School Curriculum Guide (Teaching Research), 3*, 2–5.

Dewey, J., & Ren, Z. (2005). *Interest and effort in education*. People's Publishing Press.

Herbart, J., & Li, S. (2010). *General Pedagogy*. Beijing Normal University Press.

Leslie, C., & Ross, C. (2008). *Notes on Nature*. East China Normal University Press.

Li, T., Song, Y., Cui, B., & Hou, Y. (2017). A Study on the Status of "Nature Deficit Disorder" among Youth in Qingdao and Countermeasures. *Chinese Science and Technology Journal Database (Digest Edition) Social Sciences, 6*(3), 56.

Louv, R. (2010). *The last child in the forest*. Science and Technology Press.

Ministry of Education. (2000). Full-time general senior high school curriculum plan (experimental revision). *Academic Education, 3*, 2–5.

Ministry of Education. (2001a). Basic Education Curriculum Reform Outline (for trial implementation). *People's Education, 9*, 17–19.

Ministry of Education. (2001b). Guidelines for the implementation of research-based learning in general high schools (for trial implementation). *Primary and Secondary School Management, 7*, 4–9.

Ministry of Education. (2010). Outline of the national medium and long-term education reform and development plan (2010–2020). *China Ethnic Education, 8*(3), 1–17.

Wu, H. (2018). *Research on the development and implementation of "natural notes" micro curriculum in primary school Chinese* [Unpublished master's thesis]. Sichuan Normal University.

Xu, J. (2017). Prioritizing the development of education to do a good job of education to the satisfaction of the people. *Education in Rural China, 11*, 1.

Xu, Q. (2018). Hongqiao middle school: The power of "Natural Notes". *Shanghai Education, 30*, 56–57.

Youth Pioneer News. (n.d.). *Wechat public account: ziranbijidasai*. Chongqing, China

Yu, L., Wu, X., & Liu, S. (2010, March 11). Who made Wang Ting speechless – "Scientists are ignored by teenagers" has become a topic of concern for the delegates. *Youth News*. http://zqb.cyol.com/content/2010-03/11/content_3126791.htm.

Zhang, Y. (2012). The rainy season is coming, how to avoid the rainy weather syndrome? *Shanghai Branch Life, 3*(1), 56–57.

Zhao, W., Wang, J., & Li, J. (2018). Study on the current situation of urban rivers and environment under the river head system based on GNSS – a case study of ma'anxi wetland in Beibei District of Chongqing. *Nature magazine, 7*, 32–34.

CHAPTER 4

Different Cultures, Working Towards the Same Goals

Fei Tao, Sonia Ho, Yangxue Ou, Ruihan Wang, Yan Tian, George Zhou, and Yuanrong Li

The nature notes reciprocal learning program was designed by the science research team under the China–Canada Reciprocal Learning Program framework. Both schools believed that this program could combine science and art in literacy, as it encourages students to experience nature and to conduct nature observations. By doing so, students could learn specific skills to develop a sensitivity toward nature, promote their awareness in environmental protection, and understand the human–nature relationships. The nature observation period was chosen based on the 24 solar

F. Tao • Y. Ou • R. Wang • Y. Tian • Y. Li
Southwest University, Chongqing, China
e-mail: hxliyr@swu.edu.cn

S. Ho • G. Zhou (✉)
University of Windsor, Windsor, Canada
e-mail: gzhou@uwindsor.ca

G. Zhou et al. (eds.), *Science Education and International Cross-Cultural Reciprocal Learning*, Intercultural Reciprocal Learning in Chinese and Western Education,
https://doi.org/10.1007/978-3-031-17157-4_4

terms, and it helped both sides to communicate their findings in a specific time frame.

The nature notes reciprocal learning program was officially launched in 2015 with the assistance from Canadian and Chinese graduate assistants in the science research team. The first video conference was conducted in December and involved participants from both schools. The vice-principal from the Canadian elementary school stated that their teachers and students had expressed great interest in the project. They hoped that the project would help improve mutual relationships and students' learning. During this conference, six Canadian students demonstrated their practical work, including their weather observations and data documentation. They also carried out some scientific inquiry activities. For instance, they observed the plans, researched, and then made a plane model. They also worked together to make a water filter model. After the students shared, teachers from both sides discussed how they should implement the program. They decided to set December 7, the solar term of Major Snow, as the starting date for the implementation of the program. Students from both sides conducted nature observation centered on the weather of both countries, how plants grew and how animals migrated, and the human–nature relationship. Students were encouraged to use photos, drawings, and words to record their observations. Figure 4.1 shows the timeline of how often they communicated with each other, followed by a detailed discussion of communication.

During the three years of student work exchange, graduate research assistants accounted for the majority of the task force. For example, on the Canadian side, graduate assistants regularly went to the sister school and facilitated teachers and students in carrying out their projects. Graduate assistants were also in charge of data collections, such as recording students' project activities and their final products. The science research team noticed that the ways students presented their work changed over time, from a relatively simple format at the beginning of the project to a more diverse and creative format as the project progressed. The level of student engagement also demonstrated some changes, from the pure observation of nature to reflections on the human connection with nature. We will use three cases below to illustrate the procedure of exchange and student learning from the program.

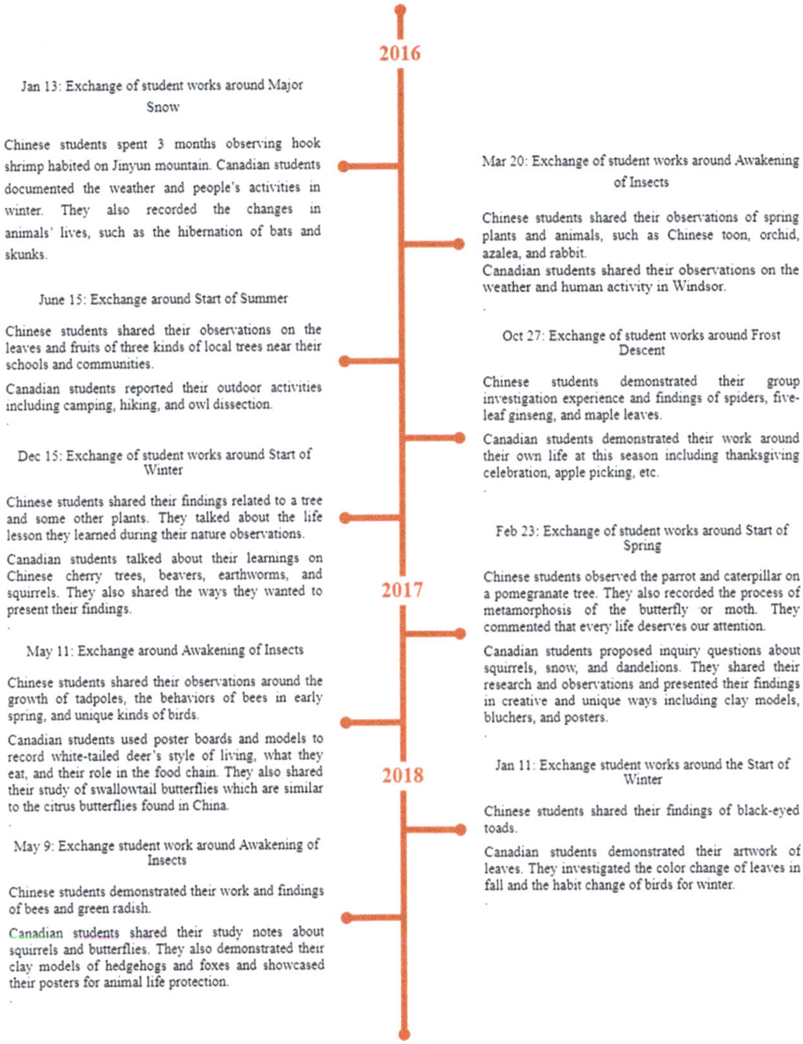

Fig. 4.1 Communication Timeline

Case 1: The First Exchange

In January 2016, the first Skype meeting took place and it was attended by students and teachers from both the Chinese and the Canadian sides. At the beginning of the meeting, a science teacher from China shared how the nature note program was implemented in their school., This was a three-step process:

1. Students first chose a specific animal or plant on which to conduct their research.
2. Students then researched their selected subjects, mainly through observations.
3. Students used photos, drawings, and texts to document their findings, as well as any inspiration that was generated when they were doing the project.

In this exchange meeting, the Chinese students shared their study findings on the hook shrimp found on Jinyun mountain. They observed the shrimp over a period of three months and found that they reproduce continually during winter. The Chinese students brought the shrimp to the lab and planned to continue to observe the shrimp in the lab environment, which can be seen in Fig. 4.2.

Fig. 4.2 Chinese students sharing their study subject (hook shrimp) in a beaker

On the Canadian side, the work of the Grade 7 and 8 students was to record the weather at different times around Major Snow. Their record also included temperature, wind strength, humidity, and snowfall. Animals' behavioral changes, including that of bats, bears, and skunks, were recorded during the season. They also reported people's winter entertainment activities, such as hockey and skiing. The students mainly used pictures to record what they observed, and then they put them on a poster board with brief explanatory text.

This first virtual exchange conference was a promising start. The Canadian teachers and students learned how the well-established program in China was carried out. They were impressed by the diligent manner in which Chinese students studied their selected subjects. Over the course of their studies, Chinese students learned how animals behave and what the community life looks like during the season around Major Snow in Canada. What Canadian students shared at the meeting inspired Chinese students and teachers to consider the inclusion of community life as a potential topic for their nature note projects in China. It is clear that the first virtual exchange conference provided solid information for the collaborative program to move forward with a better understanding of its format and product. Both sides agreed that this created a new way of cross-culture learning in science education.

Case 2: New Breakthrough in the Exchange

After the first year of exploring how to conduct the project, both sides achieved a breakthrough in the 2017–18 school year. In the Skype meeting which took place in February 2017, the Canadian students reported their findings related to squirrels, snow, and dandelions. Each student raised an inquiry question about the topic they chose. They conducted observations and also did online research. The solar term of this exchange took place around the start of winter. The student presentations of their works were very different from the first exchange. On top of pictures and text, Canadian students presented their works in many creative ways, including clay models, poster boards, and brochures, to name just a few.

One Canadian student chose to study squirrels for her project since she had noticed squirrels running around outside on a cold day. She wondered about how squirrels live throughout the year and how they keep themselves warm on cold days. In the course of her observations, she learned that there were different kinds of squirrels. She recorded what she found about how squirrels change their behavior from warm to cold days, what they eat, and how they look. She was particularly interested in the behavior of ground squirrels. She reported this kind of squirrel lives in the ground, such as in burrows and tunnels, instead of living on trees. Figure 4.3 shows how she made a clay model of the ground squirrel and summarized some essential findings on the flags beside the clay model.

In another project, a Grade 8 student studied snow. Her inquiry question was into how the snow was formed. She learned that tiny crystals form when the temperature is low and the air has a high moisture level. These ice crystals stick together in clouds to become snowflakes. When enough ice crystals stick together, they become heavy enough to fall to the ground. It was amazing to hear that no two snowflakes were alike, and that the largest snowflakes could be several inches in size. Contradictory to her intuitive knowledge, she observed that snow is not white but translucent. The student used a brochure to record the facts she collected about

Fig. 4.3 A clay model of a squirrel

snow (Fig. 4.4). She drew small pictures on the brochure and wrote snow facts in different colours. At the back of the brochure, she also listed many websites where students could find more information about snow. Through her presentation, Chinese students learned a great deal about snow, which they were unfamiliar with due to the rarity of snowfalls in their city, Chongqing, China.

Another Grade 8 student studied the dandelion, a very common plant in Canada. Her inquiry question was: How is a dandelion's lifecycle different from other plants? She realized that there were two different kinds of

Fig. 4.4 A Brochure about Snow

dandelion in the world. One is yellow, whereas the other is white and puffy. Despite the difference in color, she reported that dandelions could be used in many different ways. People eat it as a vegetable and use it to make wine. Dandelion is also one kind of Chinese herb medicine. The students found that the lifecycle of the dandelion has three stages: germination, flowering, and reproduction. Symbolically, a dandelion can represent three things: the yellow flower head represents the sun; the puffball represents the moon; and the dispersing seeds represent the star. Figure 4.5 shows how the student summarized her learnings on a poster board and shared them with Chinese students during the Skype meeting. A poster board is a popular way that Canadian students like to choose to present their findings as it has enough space for detailed information. It is also easier for readers to understand the basic concept in a clear and visualized format.

On the Chinese side, three chosen topics were caterpillar, parrot, and cicada. The first presenter was a Grade 7 student, who observed a

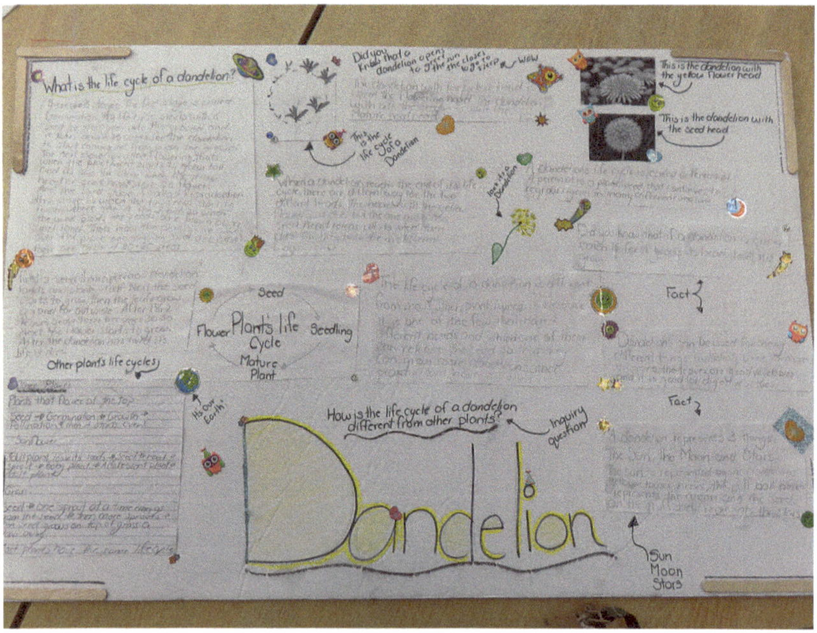

Fig. 4.5 A Poster Board about Dandelion

caterpillar on a pomegranate tree. The day after the Clear and Bright Day, this student's family bought a pot of pomegranate flowers. Everyone was attracted by the beauty of the flowers, but they did not notice a little hidden caterpillar. After the Grain Rain day, the pomegranate flowers continued to bloom in succession. She still had not noticed the little caterpillar that had crept beneath a piece of leaf. After Grain Bud's day, she went to the balcony to conduct her regular observations and found that a pomegranate fruit had been attacked by the caterpillar and was bruised. With her aching heart for the bruised fruit, she reached out, attempting to capture the caterpillar, but the pomegranate fruit fell to the ground along with the caterpillar. She treated the pomegranate fruit as a gift for the caterpillar. Before the White Dew day, the only pomegranate fruit grew as big as a fist under close attention from her family. It clung to the trunk firmly and was never attacked by caterpillars. The student reported her observation of the flowers through time and shared her inspiration from the process (Fig. 4.6). She commented that people tend to be fascinated by the fancy things surrounding them, but ignore some negligible existence; for example, the tiny caterpillar in this story, which is also nature's wonderful creation, deserves our attention.

The second Chinese presenter was a Grade 7 student who observed a couple of parrots. With persistence, the student persuaded her mother to buy her a pair of little parrots as her birthday present, which she kept in a cage. One day, she came back home from school and was stunned by the scene that she found. In the cage, there was only one parrot left. The parrot pecked at the door without being able to open it. The other parrot, which had somehow managed to escape, was still outside the cage. However, it did not fly away for freedom. Instead, the escaped parrot stayed with its companion outside the cage. The student was so touched by the scene that she decided to open the cage and set the other parrot free. Together, the two parrots spread their wings and flew away.

At the virtual exchange event, the student shared a life lesson she had learned from her project. She learned that every creature of nature has feelings, and sometimes these are even richer than those of human beings. The pair of parrots reminded the presenter that, as human beings, people should stick together after setting up families. Husband and wife, as well as parents and children, should not desert each other. However, in the present day, people consistently witness betrayal and desertion for a variety of reasons. At this point, she thought that humans could not even compare with parrots. The presenter was proud of her parrots, who would give

Fig. 4.6 Pomegranate and Caterpillar

up their freedom to be loyal to each other. She wished her little parrots would live happily ever after. Figure 4.7 shows her findings displayed in a poster.

A Grade 8 student was the third presenter in the Skype meeting. Her topic was a cicada. She started her observations very early in the morning and her observations were recorded in a detailed way (Fig. 4.8). The following section is a detailed list of her timed observations:

Fig. 4.7 Never-Separated Parrots

At 5:30 a.m., a baby cicada emerged from underneath the ground.

At 6:07 a.m., the cicada landed on a tree trunk after multiple attempts.

One minute later, the shell of the cicada cracked from the middle on its back, revealing the green flesh inside.

At 6:09 a.m., the cicada's head and mouth came out of the shell.

At 6:16 a.m., most of the cicada's body came out.

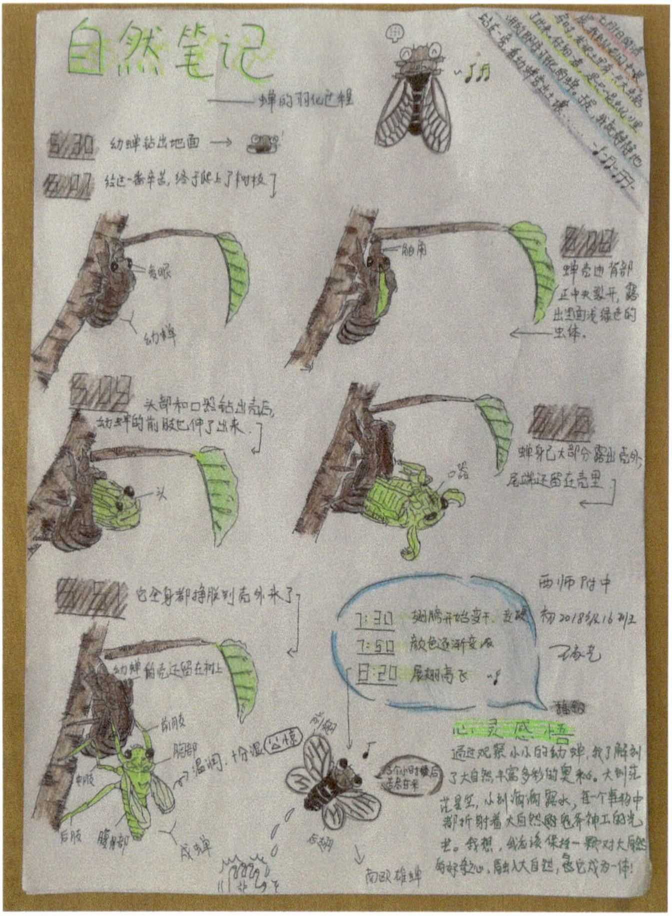

Fig. 4.8 Cicada Shedding Its Shell

About 25 mins later, the cicada completely came out from the shell and left the shell on the tree. The wings were wet at first and then gradually dried up. The cicada finally lay on a branch.

The presenter also shared her findings about one particular kind of cicada found in France. If people touched the cicada, it would excrete urine. Finally, this student shared the inspiration she had learned from the project. Through observation, she understood the power of nature, which inspired her to always hold respect toward nature. She also understood that human beings should always hold a sense of curiosity toward nature and being part of it.

In this Skype meeting, students held wide-ranging discussions about each other's projects. They raised questions after each presentation, and those questions were answered not only by the students but also by their teachers. Chinese students not only observed their study subjects, but also built their personal connections with nature. They shared their life lesson learned from their projects, which made a great impression of the Canadian teachers. During the meeting Canadian teachers expressed their intention to build such reflective components into their future project assignments. Canadian students presented their projects in innovative formats, which impressed the Chinese students and teachers. Their study approaches also revealed some differences. Chinese students mostly finished their projects solely with observations and Canadian students finished their inquiry projects with both observations and a literature search.

Case 3: Exchange around Awakening of Insects

Awakening of Insects is a solar term representing the fact that spring is back, and is the period when animals and insects start awakening from hibernation. In May 2017, Chinese and Canadian students shared the projects they did from the time of Awakening of Insects to Spring Equinox. They exchanged their observations of the nature, the information search, and the lessons learned from their projects. Three Chinese students shared their projects entitled the Growth of Tadpoles, the Sense of Spring, and the Awakening of Bees, respectively.

One Chinese student caught some toad tadpoles from the field and brought them home. He observed them and wrote a weekly log, recording their growth into toads, which can be seen below. Following the log, Fig. 4.9 displays the growth of tadpoles in a poster created by the student.

> March 5. Cloudy (Awakening of Insects)
>
> I went to the field to catch some toad tadpoles. They have a black and flat body and a long and thin tail. Numerous tadpoles gathered between plants in the shallow water. They have the size of soybeans. These little lives seemed to enjoy the bright and warm spring very much.
>
> March 12. Sunny

Fig. 4.9 Beautiful Life: The Growth of Tadpoles

The little tadpoles have grown to the size of my nail, and they are still very black. They seem to be very energetic. Their tails constantly swing when they are eating.

March 19. Sunny (one day before Vernal Equinox)

The tadpoles have grown even bigger. Now I can see their black and bright eyes and their tiny rear legs of the size of a needle tip. Their bodies start to turn into a translucent greyish-brown.

March 26. Sunny

Their bodies have grown longer now but they are still quite flat and thin. Their rear legs become curved, and front legs begin to appear.

April 3. Sunny (Clear and Bright)

I went hiking today and saw a small toad in the grass. It was yellowish-brown and as big as a chicken egg. There were many bumps on its skin. Although it looked ugly, I learned from the internet that toad has various medical usages in Chinese medicine. Toads can be used for detoxification, dehumidification, and reduction of fever.

Another Chinese student observed different kinds of birds. She also noticed that more and more insects came out. The explorations of nature made her feel that the whole world was vibrant and full of life. She wrote a weekly log to record her observations, which can be seen below.

March 5. Cloudy (Awakening of Insects)

After a spring rain shower, my dad and I walked to the hillside behind our neighborhood. The road was muddy after rain. We did not see many animals except for different kinds of birds. They were unique from each other. One kind of bird was bigger with a white face. Another kind of bird had a white crown. We also found some beetles.

March 12. Cloudy

It was another rainy day. The plants and animals I saw today were similar to those I saw last week. I saw more insects today. Some of the insects looked like flies but their wings were yellow. I also found snails on vegetable plants. Swallows appeared in our community.

March 19. Sunny (one day before Vernal Equinox)

Today is a sun-shine day. I saw a lot more insects and animals such as bees, butterflies, spiders, and ants. They were all very busy and energetic. Flowers start to blossom and grass gets green. The whole nature gets full of lives. I found an amazing bird: an arctic spirit bird. I learned that its feathers can change colors with season changes.

At the end of this presentation, Canadian students raised some questions. They wanted to know why the bird's feathers changed colors in different seasons. The Chinese students answered that it might be because of changes in the temperature. The color changes of feathers at different seasons are very commonly observed among birds.

Instead of the animals and plants in the natural context, another Chinese student investigated the life activities of honey bees at this time of year. On the hillside behind his neighborhood, there were three bee boxes that had been there for a while. Bees flew in and out every day and the student recorded his observations in a weekly log, which can be seen below.

March 15. 9°C. Light rain, Baibei district, Chongqing

It was drizzling today and the temperature was low. The bees were inactive. There were only a few of them gathered around the exit of the bee boxes.

March 15. Night.

Many bees gathered around the exit of each bee box. They clustered together and spent the night outside of the boxes. I guess it was because they reproduced too much during this period, and therefore, there was not enough room for them to stay inside the boxes.

March 16. 12°C. Cloudy

The rain stopped and the bees were active. Although it was still cold, some bees flew out and into the boxes. I could see flower pollens stuck on their legs. Apparently, these bees had started to suck nectar from flowers.

> March 16. Night.
> More bees clustered outside of the boxes than the previous night. It could be because of the increased temperature and even more reproduction.
> March 17. 14°C. Cloudy
> The temperature rose today, but it was still cloudy. The bees were even more active, and many flew in and out. Someone threw a chunk of beeswax beside the boxes, and lots of bees landed on it. Maybe they were collecting leftover honey from it.
> March 17. Night.
> The number of bees increased. More bees clustered outside the boxes for the night. They completely covered the exits.
> March 18. 17°C. Partially Cloudy
> The sun finally appeared in the sky. All bees were super active. Lots of bees gathered near the exits of the boxes. The flying in and out bees formed a bee cloud in front of the box entrances.
> March 18. Night.
> More bees had to stay outside of the boxes, and they landed around the exits. Today was my last day of observation. I do not know if there will be more bees staying outside of the boxes in the following days.

At the end of his presentation, the student concluded that, at the time of Awakening of Insects, the days are getting warmer and there is more rain. The number of bees increased and they become very active.

As mentioned above, Canadian students loved to present their findings in very creative ways. At this exchange meeting, a group of students used a poster and a 3D model to present the white-tailed deer's living style and eating habits (Fig. 4.10). They also showed their findings on the

Fig. 4.10 White-tailed Deer poster and model

white-tailed deer's role in the food chain. Their inquiry question was: What role does the white-tailed deer play in the food chain? In particular, white-tailed deer were chosen because their white tails make them very distinctive. The white-tailed deer are herbivorous animals. They eat whatever plants are available to them. In winter, when there is no grass available, the deer eat the leaves of the evergreen trees. In the eastern forests of Canada, the deer eat bugs and bush leaves. They have also learned that the deer eats twice a day: once at dawn and again in the evening. The deer's predators include wolves, mountain lions, bears, coyotes, jaguars, and humans. They have good sight, and their sense of smell is particularly acute. These qualities help them avoid predators. This group of students also learned that overpopulation is a big problem for deer, which causes them to die. Another challenge deer face is the loss of habitat due to overdevelopment by human beings. Hunting is also a danger deer have to avoid. This group of students made a poster and a clay model of a white-tailed deer.

Chinese students had never seen white-tailed deer in China. They were very interested in the 3D model and the information Canadian students shared about the deer. They asked many questions about white-tailed deer and how Canadian students did the project. The following are examples of the questions that were asked: What material did the Canadian students use to make the model? Can they see the white-tailed deer in Canada? Where did the Canadian students get this interesting information from? One Chinese student wondered if the Canadian students really went to the forests in the East to observe those white-tailed deer. The Canadian students answered those questions patiently. The exchange became an active scene in which communication, sharing, learning, curiosity, appreciation, and gracefulness merged into a gathering that crossed boundaries, cultures, and nations.

Another group of Canadian students studied the giant swallowtail butterfly in Canada. They learned that these butterflies are the largest species to be found in Canada. They have very dark brown wings with yellow spots. Their wings can grow up to a length of 6 inches. They mainly eat citrus trees and other small plants. Because of this eating habit, they have the nickname orange dogs. Some species of swallowtail butterflies live in Canada and the US, mainly in Florida, which has many citrus trees. This group of students also learned that the life cycle of this giant butterfly comprises multiple stages. The cycle starts with an egg, turns into a caterpillar, and later a chrysalis. A young butterfly then emerges from the chrysalis and eventually becomes a full-sized butterfly. The adult butterfly mates to start a new cycle. Butterfly eggs are usually orange in colour and become darker before they hatch. Caterpillars are brown and have bumps on their skins, making the caterpillar look like bird poo, which helps prevent the caterpillars from being eaten by predators. When caterpillars are in danger, they send out a foul-smelling substance from their heads, which looks like a snake's tongue. Once the caterpillar becomes older, it attaches to something and grows into a cocoon, and then metamorphoses into a butterfly.

Benefiting from the last exchange meeting, this group of students moved beyond the scientific research of their study subject. They reflected on what they learned and made a personal connection with their study subject, as was usual for Chinese students. Even though the caterpillar looks ugly like a bird poo, it turns into a beautiful butterfly. The life lesson Canadian students learned was to not judge people by what they look like because they may have a beautiful personality. Do not judge a thing from the outside because the inside might be different. From doing this project, this group of Canadian students became inspired to not judge others simply based on how they look.

To present their findings, this group of Canadian students created a giant butterfly diorama and wrote cool facts about the butterfly under the wings (Fig. 4.11). They also made a booklet to present more details about this particular animal (Fig. 4.12). The Chinese students were interested in this topic, and were eager to ask questions. They were concerned about any serious impacts that this butterfly has on the citrus trees and wondered how this butterfly could grow into such a large size. The Canadian students tried their best to answer their questions, even though their knowledge about the butterfly was still limited.

Summary and Conclusion

The exchange mentioned above indicate that participating Canadian and Chinese students learned a great deal from their participation in the nature notes reciprocal learning program. They were curious about nature and recorded and shared what they observed about nature. In this process, students built their knowledge about nature, learned different ways to conduct the project and disseminate their study results, and more importantly, built their connection with nature. The program allowed for both sides to move forward with their investigation of nature. For example, by

Fig. 4.11 Giant Swallowtail Butterfly Folding Poster

communicating with the Chinese partners, the Canadian students found that the swallow-tailed butterflies share similarities with the citrus papilio in China. They both live on citrus trees and have similar styles of living. Considering that citrus trees are both invasive species in Canada and China from the Mediterranean area, participating students raised a potential investigation task: find out the possible relationship between these two kinds of butterflies.

The exchange events between two sister schools demonstrated some differences in students' ways of presentation. The Chinese students used traditional ways, such as drawing and writing, to show their findings. Their Canadian partners appreciated their talent in drawing. Canadian students were more creative presenters and they never limited themselves in the ways in which they demonstrated their work. They used many different ways to present their work, such as brochures, poster boards, dioramas, clay and foam models, PowerPoint presentations, and booklets. One student even planned to write a song to present his findings. All these creative

Fig. 4.12 Giant Swallowtail Butterfly Booklet

forms of presentation demonstrated by the Canadian students indicated that the Canadian way of conducting their projects was full of flexibility and openness. Such differences opened up a space for both sides to appreciate different ways of learning.

In conclusion, through the nature notes program, students from both sides have developed a broader knowledge and a deeper understanding of one another's culture. They both showed a positive attitude toward their international partners' culture and values. What they have achieved from their participation had met the goals the program was designed for. This finding is similar to the results of an international collaborative educational program with a focus on scientific learning where students showed higher awareness and understanding toward the culture of their counterparts (Gillespie & Melber, 2014). Understanding each other's differences is beneficial as it helped equip the student with an open attitude so that they can have the ability to sense each other's strengths and understand each other's limitations. Developing such a mindset will ultimately benefit the students because they will have more chances to meet with people from different cultures in the globalized world.

Limitations and Suggestions

One limitation that the Canadian students faced was they could make fewer nature observations due to the weather conditions. A Canadian student indicated that "the season changes a lot in China, they have more seasons than us... They have a lot more species than we have here." Canada is a lot colder than China, and the winter lasts around six months, so the students were limited to find animals or plants to observe in that period. The Canadian students tried, but were unable to find any insects or animals in the cold schoolyard. The other limitation for the Canadian students was a lack of chances to observe nature, as taking students to observe nature depends on their parents' availability. According to one of the Canadian teachers, getting parents involved in their school community could prove very challenging. The school is located in a low socio-economic area, meaning that many of the parents needed to work and might be unable to spend much time taking children to nature. This limitation explains why many students had to search for information from the internet instead of collecting data from nature.

The Chinese students conducted multiple nature observations and accumulated a large amount of data. They felt that the most challenging part of doing this project was finding a way to demonstrate all their findings in a limited space on an A4 paper, which was the project's requirement. It became a challenge for the students as they had to select the most representative data to show their learnings. In addition, according to two students' interviews, they believed that choosing an appropriate way to demonstrate their work was another difficulty. Drawing pictures and writing data on paper were typical ways of demonstration. However, the Canadian students showed them different solutions. According to the Chinese lead science teacher's sharing in the one-to-one interview, he indicated that the time zone difference and the internet problems created obstacles to the video conferences. Internet problems affected the quality of communication and the time difference influenced the length of meetings. Hence, teachers and students were finding it hard to have smooth interactions. These were the major challenges faced by both sides. Data show that the low quality of communication caused by the internet speed and the time difference hinders effective interactions between the two schools. However, a clear and effective interaction is required when the two sides come up with ideas while conducting a collaborative activity. According to Yang et al. (2014), who studied the students from China and the US, they found that those students tend to have better in-depth individual communications, and build more prior knowledge of each other's cultures. In applying their findings to this study, the students from both sides can exchange ideas on similar topics. They can form one-to-one dialogue or small group chat for developing in-depth communication. Both settings allowed members to communicate their thoughts deeply in the virtual space without distracting others. Setting up in-depth communications will have different benefits for students. O'Neill (2007) found that the students' social interaction skills, writing, and reading skills can also be developed. The asynchronous online communications in a content management system (CMS) helps students develop culturally meaningful communications because it promotes students' intercultural competence and their motivation for learning.

References

Gillespie, K. L., & Melber, L. M. (2014). Connecting students around the world through a collaborative museum education program. *Journal of Museum Education, 39*(1), 108–120. https://doi.org/10.1080/10598650.2014.11510800

O'Neill, E. J. (2007). Implementing international virtual elementary classroom activities for public school students in the U.S. and Korea. *Electronic Journal of e-Learning, 5*(3), 207–218. Retrieved from https://files.eric.ed.gov/fulltext/EJ1098758.pdf

Yang, J., Kinshuk, Y., Chen, S. J., & Huang, R. (2014). Strategies for smooth and effective cross-cultural online collaborative learning. *Journal of Educational Technology & Society, 17*(3), 208–221. https://www.proquest.com/scholarly-journals/strategies-smooth-effective-cross-cultural-online/docview/1556986520/se-2

CHAPTER 5

Leading the Way for Students' Growth

Sonia Ho, Jin Mu, Yiling Qi, Liujun Guo, Shijie Deng, George Zhou, and Yuanrong Li

Xingzhi Tao (1891–1946), a famous Chinese educator who studied in the USA in the 1930s and whose work was strongly influenced by John Dewey, said: "Action initiates knowing, and knowledge results from action" (Tao, 1934/2016). After the nature notes program was introduced in the Chinese schools, students began to walk out of the classroom and step into nature. Striving for guided participation in practical outdoor activities under teachers' guidance, the students gradually improved their abilities to observe and feel things in nature around them. While observing nature, not only can the students learn knowledge but they can also develop an understanding of their relationship with nature. In this way,

S. Ho • G. Zhou (✉)
University of Windsor, Windsor, Canada
e-mail: gzhou@uwindsor.ca

J. Mu • Y. Qi • L. Guo • S. Deng • Y. Li
Southwest University, Chongqing, China
e-mail: hxliyr@swu.edu.cn

G. Zhou et al. (eds.), *Science Education and International Cross-Cultural Reciprocal Learning*, Intercultural Reciprocal Learning in Chinese and Western Education,
https://doi.org/10.1007/978-3-031-17157-4_5

students experienced how to respect nature and learned how to get along with nature.

The nature notes program has been held consecutively for seven years from 2014 to 2021. Over the seven years, the development of students' nature notes has transformed from tender to mature. The first participants of the program have entered high schools and colleges with the seeds of their dreams planted at a young age, having grown roots and sprouts. The following section includes detailed stories of the students who took part in the program.

Student Stories with the Nature Notes Reciprocal Learning Program

Student Story 1: Protecting the Mother River—The Youngest River Captain in Chongqing

Chongqing is a city located in the southwest hinterland of China, nicknamed Jiangcheng (river city). The city is located in the upper reaches of the Yangtze river, with its mainstream cutting across the whole territory from the west to the east. It runs up to 665 kilometers, cutting across the three back slopes of Mountain Wu and forming the world famous three gorges: Wutang Gorge, Wu Gorge, and Xiling Gorge. Jialing river coming from the northwest, zigzags threefold through the city center into the Yangtze River. Wu river joins Yangtze river in the Fuling area, forming Libi Gorge, Wentang Gorge, and Guanyin Gorge (three gorges on Wu river).

ZDT was the class monitor and the president of the Water Love club in the Chinese sister school. She had an impressive title, Jialing River-Saddle Creek Stream volunteer captain. She was the youngest volunteer captain officially identified in Chongqing. This title of the caption reflects her participation in the nature notes program. In grade 7, ZDT learned about the program from her teacher and knew that the school was one of the 30 Young River Captain and Young Lake Captain pilot schools across China. She self-introduced herself to the project and wanted to take action to protect the mother river (Yangtze River). She formed a group of student volunteers and carried out a series of voluntary activities. She said: "I grew up by the Yangtze River side and have a deep feeling for it. I consider the protection of the river as a part of my life. The title of Young River Caption represents responsibility and accountability. I want it to be more than an empty title. I want to do something for it."

ZDT discovered a new frog species living in Chongqing: a white-jawed pan-tree frog. There was a glass observation tank at her home, which she and her father had built together. Inside the tank were potted plants and an artificial mountain with sporadically planted greenery. This little tank carries her stories with the nature notes program. During our interview with ZDT, she spoke with confidence about the tank contents, just like counting house gems: "This clover is planted when I did the experiment of symbiotic nodules as a result of the co-living between root tumor bacteria and legume plants. Don't be tricked by the appearance of the small shallow fish-free pool; in fact, there is a tree frog, which is the one I found as a new record of the species." Proper nouns and special terms bounce out one by one. Compared with an average junior high school student, the 13-year-old ZDT was slightly different.

Out of the school's back door, the Young River Caption, ZDT, countlessly walked on this section of the slate road to the Saddle Creek stream. Patrolling the little river became a routine on weekends. ZDT examined and recorded soil erosion in the fallen zone during each river patrol. She also carried a bag and picked up trash along the way. She wrote down the inspirations to protect the river in her notebook (Fig. 5.1). Carrying the volunteer team flag, with pliers and bags in hands, ZDT and a dozen red-vested team members picked up trash along the beach. ZDT said, "regardless of the sizes, we have almost walked over all the rivers in BeiPei district. If we find that there are some people fishing, picnicking, or growing vegetables in the drop belt, we will communicate with them and stop them from doing what they were doing." Each patrol lasted about two hours and took place once a week. The 13-year-old girl ZDT adhered to this routine for more than a year.

When talking about river conservation, the non-stoppable ZDT had no intention of pressing the pause button. "The purpose of discovering new species or new species records is to protect them," ZDT said. "Once found, the creatures will have one more documented habitat in Chongqing. We should try our best to protect them from getting harmed." "Each patrol trip is a hard journey." ZDT continued, [for example] "during the National Day Holidays, the volunteer team went out for five days. We stayed in the hotel during the daytime to catch up with our school homework. At around 7 pm, we went on the patrol trip with the teacher. We often returned to the hotel at around two or three o'clock in the morning."

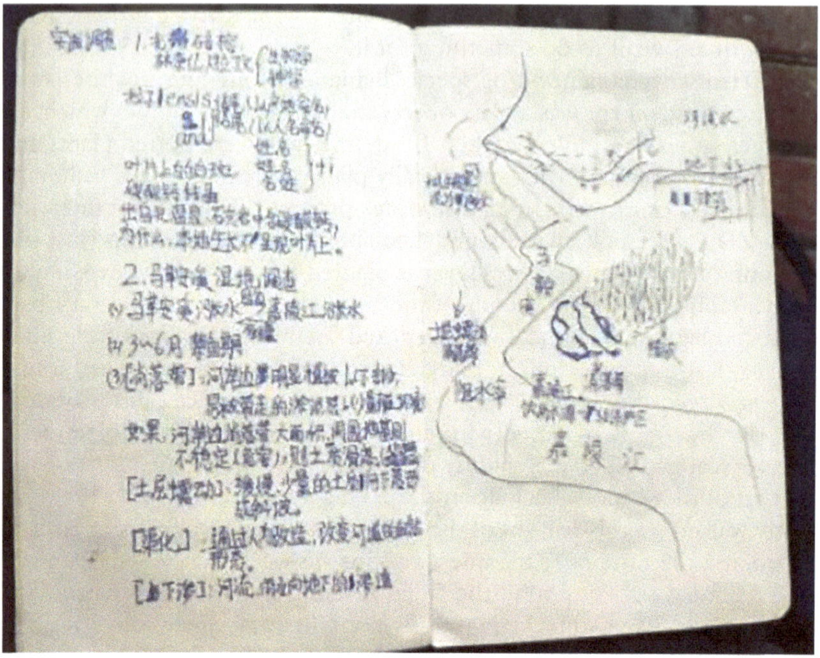

Fig. 5.1 ZDT's Drop Belt Notes

It turned out that since last year, the team and their mentor teacher went on field trips together to carry out natural expeditions. Before each trip, the team determined a target species. They would travel from Mountain Four-Sided, Mountain Fairy, Mountain Zhaomu, Mountain Lushan to Yellow Water, and from the natural forest and the caves to the villages. Each year, they made more than ten of these trips, which would take more than 20 days overall.

ZDT's dad was very proud of his daughter. "Catching the colorfully patterned tree frog is a little thing to her. Catching insects to feed lizards is nothing neither to her." While talking, her dad opened some photos on the computer. "Look, this is a picture of ZDT. She was soothing a dark brown snake with her hand." ZDT glanced at it, raising the corner of her mouth, and said, "this one is not poisonous."

ZDT's achievements after these efforts were eye-opening. Her project on the application of Rhizomes in urban ecological gardens won the first prize in an international competition held on one Global Nature Day. After three months of adhering to the daily observation of tree frogs chirping, hugging, and spawning, she won the grand prize at the Chongqing 5th nature notes contest for her poster presentation of what she learned about tree frogs. Her poster presentation on leopard cats won the first prize at the nature notes contest hosted by the Ministry of Ecology and Environment.

Because of such outstanding achievements in the nature notes initiative, ZDT and the other two students became the only high school student team in the country that was interviewed by the Director of the Information Division of the United Nations Environment Program (N.d., 2019). "Mr. Nathan Saba signed for me and encouraged me to continue my inspiration in scientific inquiry and contribute to environmental protection." Speaking of this experience, ZDT said, "it is a very honorable moment. It also ensures me that the efforts of environmental protection are valuable and meaningful."

ZDT was awarded as the national Model Youngster in New Era. She was also named Jialing River-Horse Saddle stream volunteer Caption, by the Beipei District River Office. ZDT said, "I am just a very ordinary high school student, putting aside my love of natural rivers and commitment to environmental protection."

ZDT's story shows that her participation in the nature notes program was inspired by her love of the rivers where she grew up, which is her internal motivation. The encouragement from the Director of the Information Division of the United Nations Environment Program and the awards she received from nature notes contests provided her a positive boost, inspiring her to go further down the road to protecting her mother river. Implementing the nature notes program has enabled many middle school students like ZDT to experience nature, love nature, and protect nature through active engagement with nature. The program's impact on participating students takes place in a way that the spring rain influences all lives, watering them finely and silently!

Student Story 2: Practice Generates Knowledge Through Field Exploration

"The nature notes project started my journey of exploration. I wanted to offer my contribution to nature using what I learned at school." This is how HYF began to introduce her story with the nature notes program.

Shortly after entering the high school, HYF and her classmates were given a task using the format of nature notes to record the natural scenes that interested them. Born and raised on the banks of the Wuxia River, this girl, who grew up with a keen interest in nature, decided to pick a simple topic to work on: ferns.

"How naive I was proved to be! Ferns looked simple in structure, but when I started to draw it, I realized that the leaves of different ferns vary. It required careful observations to draw correctly how the compound leaves split and how the sporangia grow." HYF recalled her first nature notes project. This project won an excellence award in 2015 in the second Chongqing nature notes competition.

In the summer of 2015, HYF was about to enter her second year of high school. She took part in a field science expedition organized by her school. At Jinfo Mountain, she and her classmates, under the guidance of a lead teacher, recorded the chain snakes with a black-banded belly and the red-spotted toads for the first time in Chongqing.

When the research team interviewed HYF, she was a university student studying agricultural water engineering at Sichuan University. She often had fieldwork in remote rural areas and explored how to use the natural resources in environmentally friendly ways. HYF explained how she chose her university major, "My personal interest in nature drove me to choose to participate in the nature notes program. If I hadn't participated in the nature notes program, I wouldn't have become so close with my guide teacher. Without the encouragement from my guide teacher, I wouldn't have participated in a field trip to Jinfo Mountain, and I wouldn't have determined that I liked biology. Although my university major is not biology, but it is related to environment. I am still working on tasks on environmental protection."

In this statement, HYF expressed her growth experience in the nature notes program. In a later interview with her mentor teacher, we learned that HYF came from the countryside. She was interested in nature since she was a child. Her living experience in the countryside built her connection with nature and became her internal drive for her to explore nature. After being exposed to the nature notes program, under the professional guidance of her teachers, she not only gained more specialized knowledge, but more importantly became aware of environmental issues and took actions to protect the environment.

Student Story 3: Embrace the Nature and Build a Harmonious Parent-Child Relationship

Nature is a real and rich encyclopedia that contains great educational wealth. Parents and children can together go into nature without many limitations. As long as parents provide children guidance and accompany and watch for their safety, children can easily become good friends with nature, which not only benefits children physically and mentally but also cultivate their temperament.

GYQ was the winner of the first prize in the 5th nature notes contest in Chongqing. He was fishing with his grandfather and father when he happened to come across the Chinese sand minnow and started his own research. He described the process of his project using a couple of sentences,

> Dad caught something and he believed it was a fish loach, but grandfather said it was a steel loach. They argued. When we got back home, I went on the internet and found that it was neither a fish loach nor a steel loach. I approached a professor for help. He told me that the real name of this creature was Chinese sand loach, a unique fish species in China. I put it back into the river and let it return to nature. I also told my dad to release it in the future fishing.

The judges of the contest commented on GYQ's work this way, "a fishing trip triggered the curiosity of three generations of grandfather, father, and grandson about a fish loach. The experience-based argument between grandfather and father, son's curiosity, the use of modern technology, and the consultation of an expert finally lead to the clarity of the study topic."

What a memorable trip with family members! It was this wonderful family experience that allowed little GYQ to get to know the endemic fish species. Throughout this project, GYQ did not only learn extracurricular knowledge, but also greatly appreciated the togetherness with family members in nature. The parent-child relationship helps improve the environment of family education. Positive and effective parent-child communication can enhance the emotional connection between each other, which enhances the sense of belonging of each family member.

Rain is a very common natural phenomenon and does not usually trigger our curiosity. But LFQ's nature notes project reported a rain-related study he and his father worked on. He illustrated their study,

> I went to the roof with my father and found a lot of shiny water droplets hanging on the clothesline, how strange! I searched the internet and

> couldn't find the reason, so I decided to do an experiment with my dad to explore the formation of water droplets on the steel pipe. We poured water on the pipe first, and there were no drops left on the pipe. Then we reduced the amount of water. Finally, we realized that the shower of last night was the reason for the water droplets on the clothesline. Nature has so much for me to learn. I need to study hard at school.

The rain-related observation and follow-up series of experiments drew a vivid picture of working together between father and son. In the process of parent-child exploration of nature, parents need to pay attention to children's every move, appreciate their struggle and happiness, and respond to their various questions. Parents should have a mindset of playing with children and being a bit childish. The most enjoyable thing to children is their parents' positive mood and adequate patience. Exploring nature together will bring parents and children closer.

Student Story 4: Battling with Wisdom and Courage to Fulfill the Career Dream of Environmental Protection

"The things that make me want to know what they are and what they are all about are the things I should spend most of my time observing," ZW said calmly about her experience in the nature notes program. Her persistence and love for environmental protection affected those listening to her story.

ZW was first introduced to the nature notes program when she took the exploration-oriented curriculum in her first semester of high school. By chance, she and her classmates found a small black and yellow caterpillar on the way to school. With a desire to explore the tenacious life and a love for living things, she took the little bug back to her home to take care of it. After waiting patiently for 25 days, she observed the metamorphosis of life from worm to chrysalis to butterfly and recorded each stage with photos and videos. While observing, she recorded different questions each time, such as, "why do the little bugs change more at night?" A journey of heartfelt feelings and continuous exploration was recorded into a graphic form entitled, *Insect, Chrysalis, Butterfly Transformation Axis* (Fig. 5.2). This work won the first prize in the 4th nature notes contest.

ZW said that making nature notes often requires long hours of observation and recording. Her initial curiosity and interest were fundamental for her persistence. ZW shared her experience of participation in the nature notes program with those who were first introduced to the program. "I once spent 25 days continuously observing and recording the metamorphosis of a caterpillar. In order not to be interrupted, I sometimes even

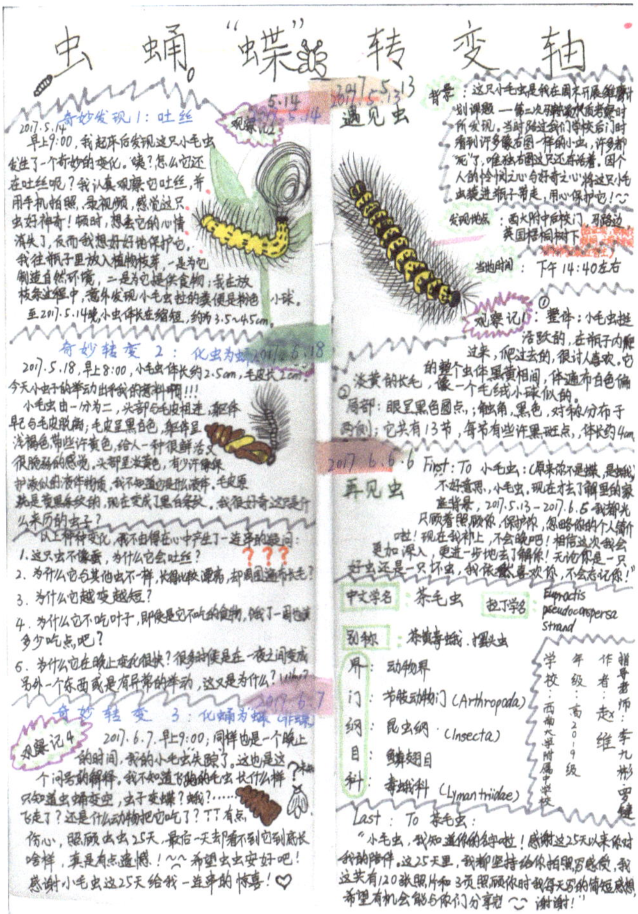

Fig. 5.2 ZW's Natural Notes Work: Worm, Dragonfly, Butterfly Shift Axis

carried that caterpillar with me, feeling that my whole being had been completely immersed in the caterpillar's world." It is this thirst for knowledge of the unknown world that made her reflect and grow along the way, constantly exploring the larger world!

Since then, ZW had been on the road of exploring and recording nature. She and her cousin jointly completed a project entitled, *New Discovery of Black-eyed Toad* (Fig. 5.3), which won the grand prize of the 4th Chongqing nature notes contest. In the closing ceremony of the contest, she said that her constant observation of nature made her have a dream of enrolling in a college majoring in environmental protection.

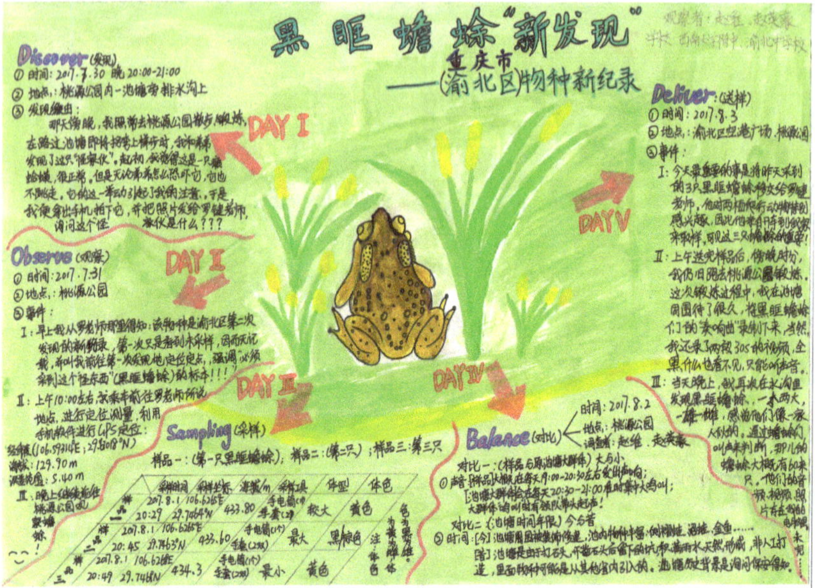

Fig. 5.3 The 4th Nature Note Contest Grand Award Winner: New Discovery about Black-framed Toad

Since her first nature notes project in her first year of high school, ZW had been fascinated with every "date" with nature. Under the guidance of her guide teacher, ZW had already published a paper on black-eyed toads and had conducted indoor breeding of black-eyed toads under the guidance of her teacher. She discovered that there was a half-hour time difference between the black-eyed toad's calls under different environments indoors and outdoors. It is understood that ZW was chosen to be a member of the Chongqing Youth Innovation Talent Development Program. In addition to her daily studies, she often went to university laboratories to participate in experiments with professors and experts.

The once high school student had now finished her three-year high school study. After graduating from high school, she chose to major in environmental engineering without a second thought. After an in-depth conversation with ZW, we found out that she loved nature from the bottom of her heart. This was why she could always stick to her original intention. She mentioned to us an episode that took place when she decided on a university major. Her parents wanted her to study economics or finance, which was more popular in the Chinese job market. However, she finally convinced her parents to allow her to study what she loves.

The nature notes program has influenced one group of students after another. ZW's initial involvement in this program was no more than a normal trial in her life. However, it was this trial that led her to the field she was interested in and pushed her to keep going deeper and deeper into this field for research. The animals and plants she met during her involvement in the program made her fall more and more in love with nature. She realized how significant the natural environment was for the survival of plants and animals and therefore she chose to major in environmental engineering. Her internal love and curiosity about nature motivated her to further study it. External factors, such as awards, reinforced her motivation to get more involved in nature study.

Student Story 5: Hanging Foot Tower as Culture Memory

Beijing's Courtyard, Shanghai's Shikumen, and Chongqing's Hanging Foot Tower are all original and distinctive urban residential buildings in China. The hanging foot tower is an ancient residential building in southwest China that has a history of 3000 years. It is the iconic cultural symbol of the mountain city, Chongqing. The wise people of the mountain city built their residences by the mountain. These buildings were often situated at the side of a hill and beside a body of water with a shape of a sitting tiger. They were decorated with a green dragon on the left, a white tiger on the right, a red sparrow in the front, and a humous at the rear. These buildings were often oriented to face the south or the north. Hong Yadong in Chongqing is a restored and reconstructed building cluster with the style of hanging foot towers. It is the witness of Chongqing's history and culture and the symbol of Chongqing's city spirit.

LJL, an elementary school student in Chongqing, first learned about the hanging foot tower from her grandmother.

> My grandmother told me that whenever the sun sets, the golden sunlight falls on the group of hanging foot towers at the area of the river gate. These buildings were at different altitudes and with different heights and sizes, but all were designed and located in a beautiful hormonic arrangement. Standing in front of these buildings, one can see in the far distance the shining weaves on the Jianling river. ... Unfortunately, these buildings were torn down. There are no handing foot towers that can be seen in urban districts.

LJL was impressed by his grandmother's vivid description of old hanging foot towers. He decided to learn more about them. LJL learned that there were some wooden hanging foot towers in the close-by small town of Zhongshan. They were protected well. LJL traveled to the town by bus and started his field research about hanging foot towers.

LJL investigated the climate characteristics, geographical location, topographical features, and local plants of Zhongshan Town, and then carefully analyzed the building site, building materials, and architectural style of the hanging foot towers. His research led to a poster presentation entitled, *Nature Notes on the Hanging Foot Towers in Zhongshan Town* (Fig. 5.4). LJL summarized what he learned from his nature notes project by stating,

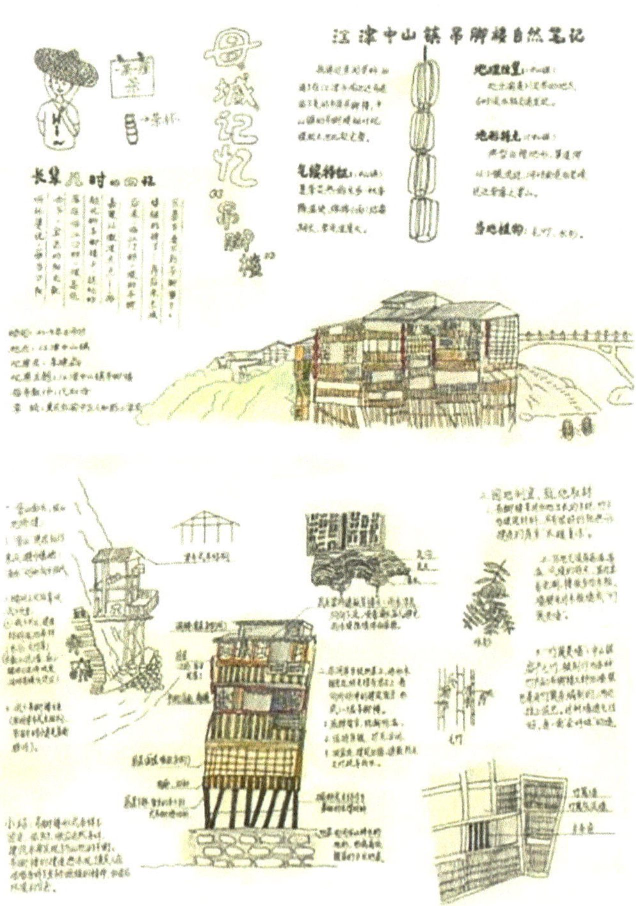

Fig. 5.4 LJL Nature Notes—Hanging Feet Towers

> the process of completing this nature notes project taught me how to observe and think about my study topic carefully and how to record what I saw and heard by handwriting and hand-drawing. I learned how the unique scale of the hanging foot towers are closely related to the local geography, climate, and vegetation. I was deeply impressed by the survival wisdom and tenacity character of Chongqing people and fully appreciate the importance of harmony co-existence between human beings and nature.

In the process of recording the nature notes of handing towers, LJL got to know the traditional cultural knowledge associated with these unique buildings through field research. He learned that the building materials came from nature, and the location of the buildings needed to be carefully chosen to fit into the geographical environment. The structure of the buildings should consider the climate characteristics. Therefore, his learning went beyond the physical feature of the hanging foot towers. Through the project, he was able to appreciate the wisdom of the local people and the charm of traditional Chinese architecture. Through the display of his nature notes, LJL not only deepened his understanding of the traditional architecture of the hanging foot towers but also made more people feel the charm and value of Chinese architectural culture through the exchange of works.

Student Story 6: Survival of the Fittest

Survival of the fittest is a law of nature and applies to human beings. WNY was one of the grand prize winners of the 3rd Chongqing nature notes contest. Her project entitled, *The Scent of Corn,* illustrated that the way that humans treat nature with kindness is not always what nature needs (Fig. 5.5). The contest judges commented on her work in the following way:

> Unlike many other middle school students who commonly used watercolor paint, the author of this product used cartoons and comics to illustrate what she wants to express in the text. The style of drawing was highly compatible with the content of the work. This work nicely recorded an interesting project of discovery. The author noticed a puppy eating corn and found that there were corn kernels in the puppy's poop. She searched for information and learned that too much feeding with inappropriate food could cause the puppy digestive problems. The most remarkable thing is that the author concluded with a reflection: are we human beings necessarily loving and protecting animals by treating them in a way we think is kind?

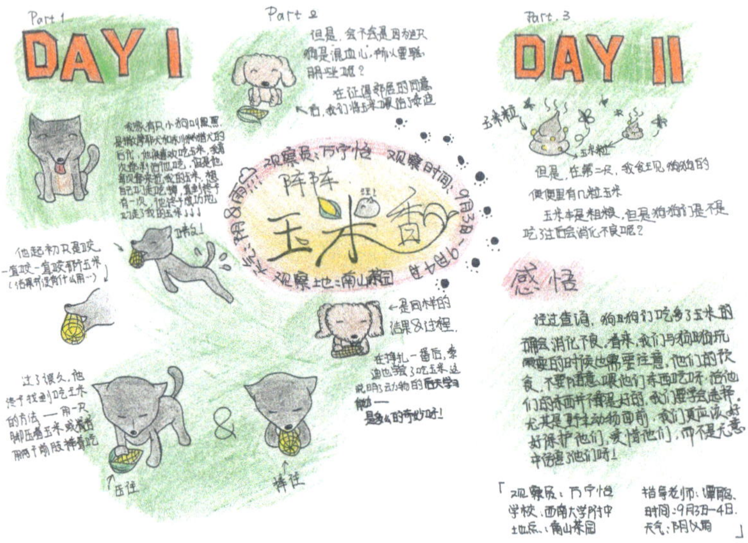

Fig. 5.5 The Smell of Corn

During the conversation with WNY, she told us another scenery she observed in her school:

> There was a newly built bird's nest outside the classroom window. My classmates were attracted by it and often watched the bird at the window. Birds were usually afraid of people, but this one was particularly brave. On a closer inspection, we found that it was a mother bird. She had laid three eggs and had been working hard to protect her children. One day we found that two eggs were hatched, but the third egg showed no sign of hatching. The mother bird led the two young birds to fly away, abandoning the third egg.

In order to provide the two young birds with a better living environment, the mother bird had made the decision to abandon her third egg. This observation provided WNY to experience the meaning of the law of survival of the fittest.

Later in the conversation, we learned that she had chosen to major in medicine at university. The medical profession has always been a popular and well-respected career choice in China, although it is a difficult program to get into and is a strenuous finish. From her initial love of chemistry and biology to her participation in the nature notes program in high school, she learned about the nature, contacted nature, observed nature, learned the laws of survival in nature, and then bravely faced the challenges and made her own professional choice. This reflects her application of the natural survival law to her own life.

Unlike the student participants mentioned earlier, WNY did not choose an environment-related major. She chose a major in medicine. Someone may think the participation in the nature notes program did not have an impact on her. However, WNY applied what she learned from nature to the human society. She believed that people need to follow the law of survival in society. She chose a more suitable and challenging career for herself. For her, the impact of the program is not only a choice of profession, but a way to help her understand the philosophy of life.

Student Story 7: The Subtle Influence Is Worth a Thousand Words

To further explore the impact of the nature notes program on the students' future careers, we interviewed the guide teacher of both HYF and ZW for their award-winning nature notes projects. The teacher commented on these two students:

> Both HYF and ZW came from rural areas. Their families were not able to provide them with professional guidance. Before they joined the high school, they had zero experience with the nature notes program and other science inquiry activities. After they came to the high school, they started to learn about environment-related majors through participating in the nature notes program. Particularly for ZW, I always encouraged her to participate in science and environment related school activities. In this process, she was particularly concerned about the ecology of the Maanxi Watershed and the Three Rivers watershed, which led her to the major in environmental protection. Based on her study of the watershed, I guided her to write two letters to the Chongqing municipal government, reflecting the current situation of the ecological environment in the Maanxi Watershed and suggestions for its management. These letters received positive feedback and timely responses from the government. Such recognition inspired her confidence in making contributions to the protection of green mountains and water.

Through the teacher's analysis, four factors can be summarized that influenced the two students to choose environmental majors:

1. An interest in nature inspired by their personal upbringing.
2. The platform provided by the school to participate in the nature notes program.
3. The subtle influence of the schoolteachers in leading students in the practical process, the charisma of the teachers' teaching, and their profound knowledge.
4. External incentives and support from the government, media, etc.

After an in-depth analysis, we use the following diagram (Fig. 5.6) to represent the relationship between the environment, internal motivation, external motivation, and students' choice to delve deeper into research.

The nature notes program stimulates and deepens internal momentum. ZW was not initially exposed to the practical activities in the environment. What she was familiar with were plants and animals that she had experienced in her daily life. After entering high school and participating in the nature notes program, she became interested in environmental issues through her exposure to the water project. With the support and encouragement of her teachers and the Chongqing municipal government, she eventually decided on her university major in the environment. From this process, we can see that the nature notes program inspired her cognitive motivation, and the continuous practical exploration and external encouragement deepened her internal motivation.

The determination of a professional career requires both internal and external motivation. ZW, who continuously recorded the process of moths from lava to cocoon to butterfly for 25 days, took nearly 130 photos. While their own internal cognitive drive pushed them to continue their

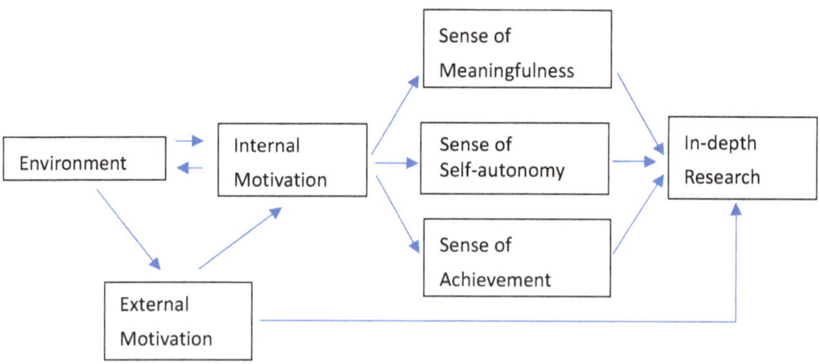

Fig. 5.6 The Internal and External Motivations for Research-based Learning

research, external feedback such as the Chongqing municipal government was an affirmation of their actions, and this external motivation was more like a "cardiac stimulant" that strengthened the direction of their professional decision.

In addition, students need to be trained in career planning through a combination of practical classes and academic classes. In an interview with teacher L, he said, "it's not the teacher who should tell students to choose what kind of career choices. Teachers should respect students' own feelings and experiences." We learned that teacher L was good at influencing students with his own actions in practice. Without dictating, he allowed students to make decisions based on their own experiences. In China, there was no formal career guidance for high school students, so teacher L suggested that a mandatory career planning course should be implemented in the first year of high school. In addition to the career guidance course, teachers should be sensitive to students' interests and provide students with opportunities to reinforce their interests and strengths through comprehensive practical activity courses.

In China, after each year's College Entrance Examination (on June 7th and 8th of each year), senior high school graduates across the country must go through the process of applying for colleges and choosing their majors. In this process, many high school graduates and parents often face two challenges. The first challenge is that they need to select a university which their exam scores qualify them for. Careful checking of the ranking of the university, the admission requirements of the university, and the quota of admission are critical factors to consider for success. The second challenge is that students must choose a promising major that maximizes the possibility of being employed after graduation. According to Tang (2013), many high school graduates do not know what they really want to study at university. The biggest disadvantage of an employability-based selection method is that students may not be interested in the major they choose to study. To make it worse, Chinese universities do not offer much freedom for students to switch majors easily.

As an old Chinese saying, "it takes ten years to grow a tree, but a hundred years to make a man." The influence of education on students is not something that happens overnight but takes place in a subtle way over a long period of time. Students spend more time in school than in any other place. School programs should not aim at knowledge transmission but facilitate the discovery of their own interests and development. It is well known that high school education in China is highly competitive and exam orientated. It is the lack of research-based learning activities, like the nature notes program, that leads students to gradually lose themselves in the heavy academic work. They do not have the opportunity to find their interests and potential, which causes difficulty in choosing a major for their higher education.

The Impact of the Nature Notes Reciprocal Learning Program on Chinese Students

Life is education and society is school (Tao, 1934). Nature is full of all kinds of insects and animals, flowers and trees, forests and wilderness, rivers and streams. The nature notes program provides students a platform to get into nature, explore its mysteries, and experience the joy of studying in nature. In the process of observing the natural environment, students not only learn scientific knowledge and exercise scientific thinking, but more importantly, they learn the methods of scientific inquiry and improve their ability to investigate things. Table 5.1 shows the awards won by the Chinese sister school in the national nature notes competitions in 2018 and 2019. Besides these student awards, the Chinese sister school was recognized as an excellent organizer in the 2018 National Adolescent Nature Note Contest and the 2019 2nd Changlong Cup National Nature Notes Contest.

The nature notes program aims to enrich students' scientific knowledge, improve their scientific research skills, develop their scientific literacy, and deepen their understanding of the impact of science and technology on society and individuals. To examine the impact of the

Table 5.1 Awards Southwest University Affiliated Secondary School students received in the past National Natural Note Contests (2018 and 2019)

Contest	*First prize*	*Second prize*	*Third prize*	*Special award*	*Merit prize*	*Mentor prize*	*Total*
2018 National Adolescent Nature Notes Contest	1	2	14	4			21
2019 National Adolescent Nature Notes Contest	5			3	3	6	17
2019 2nd Changlong Cup National Nature Notes Contest	1	2	1	4	6	17	31
Total	7	4	15	11	9	23	69

Table 5.2 Program impact on participants

	Percentage (%)				
	1	*2*	*3*	*4*	*5*
Knowledge	0	2.15	19.55	56.85	21.35
Skill	0	1.40	19.60	52.90	26.10
Self-development	0.97	4.57	23.67	45.17	25.60

nature notes program on students, the Chinese research team conducted a study on the scientific literacy and self-development of 138 students in their first and second years of the nature notes program at the Southwest University Affiliated Secondary School. In this study, knowledge refers to the students' knowledge reserve after participating in the nature notes program. Skills refer to the skills they acquired through participating in the program, including research skills, communication skills, and collaboration skills. Self-development refers to the students' acquired abilities for personal development through the program, such as career choice. The 5-point Likert scale was used in the survey with 1 to 5 representing strongly disagree, disagree, unsure, agree, and strongly agree (Table 5.2).

The survey results show that over 78% of students agreed or strongly agreed that their knowledge about nature had improved significantly, with about 20% of students not very sure. About 79% of the students agreed or strongly agreed that their inquiry skills had been improved significantly, with about 20% of students not very sure. Over 70% of the students agreed or strongly agreed that they had gained significant self-development, with 24% of students not very sure. It can be concluded that after participating in the nature notes program, students not only expanded their knowledge and improved their interpersonal skills and cooperation skills, but also gradually determined their future career goals and their life direction.

Overall, we believe that the nature notes program enriches students' knowledge, improves students' abilities, and promotes students' self-development. The following section defines each of these ideas in detail.

Enriching Students' Knowledge

Connecting with Reality and Consolidating Classroom Knowledge After participating in the nature notes program, students' knowledge is enriched. Students use the nature classroom to observe and explore plants and animals, which not only enables them to be inspired by the natural environment, but also consolidates their knowledge learned in classrooms. Students combine their theoretical knowledge with natural phenomena, learning by doing, and doing while learning. ZW measured the latitude and longitude of the observed objects with the help of the Beidou Satellite Navigation System (BDS), thus recording observation notes in more detail. The concepts of longitude and latitude were learned in the geography classroom, but by combining the knowledge learned in the classroom with practical exercises, ZW further understood the practical meaning of the knowledge and improved his ability to apply the theoretical knowledge. The combination of theory and practice helps strengthen students' interest in learning and exploration. During the observation trip to Jinyun Mountain, teacher L introduced the family, genus, and species of a pepper tree, and the common type of pepper trees students often see in their neighborhood. Such field trips provided opportunities for students to connect theoretical knowledge with nature, inspiring them to learn more. Observation and Exploration to Acquire Knowledge About Nature When working on the nature notes project, students can acquire new knowledge through their own observations or by searching for information. HYF shared her field research experience, "while conducting research in Jinfo Mountain, we found a special kind of toad called red-spotted toad in the cave, and then recorded their sound spectrum with the help of university instruments and found that we could identify their genders through the analysis of their sound spectrum." In the process of exploration, HYF not only discovered the red-spotted toad, but also got a chance to use the acoustic spectrum instrument borrowed from a local university to determine the acoustic spectrum of the animal and analyze the test results. She learned how to distinguish male and female animals by analyzing the acoustic spectroscopy, which laid the foundation for subsequent relevant research studies on other animals. Student ZW learned a large amount of knowledge about the conservation of water resources while she worked on her nature notes projects on rivers and lakes. During the interview, ZW concluded,

> I have definitely increased my knowledge of water conservation through my participation in the activities. Firstly, I learned the basics of water resources, such as the classification of water qualities, the indicators of water components, and the advantages and disadvantages of each component. Secondly,

> I learned about how to monitor water quality, such as how to measure water components and how to trace the origins of water pollutants. Thirdly, I learned knowledge about simple and feasible ways to protect water resources and the investigation of technology for water treatment. Finally, I developed a good habit of protecting water resources.

Besides the learn knowledge about nature, ZW also shared her learning of Chinese traditional culture in the interview:

> Initially, my knowledge about the 24 Solar Terms was mostly from the older generation. In my hometown, the 24 Solar Terms were rarely used, so I knew little about them. After participating in the nature notes program, I not only became more aware of the timing of each solar term, but also became more sensitive to the changes in plants and animals and more aware of the customs of the society around each solar term through the observation of the nature and human activities. Secondly, during the project, I also searched some ancient poems about the 24 Solar Terms, which expanded my knowledge about them and deepened my understanding of the traditional Chinese culture. In addition, through the exchange events, I was able to demonstrate the Chinese culture and local ecology to Canadian students and simultaneously learn the Canadian culture and society.

Improve Students' Overall Ability

Improve Observation Skills In the process of participating in the nature notes project, students went into nature, observed nature, and took the initiative to discover the mysteries of nature. In return, their observation skills greatly improved. Students became more sensitive to the changes of nature and more motivated to discover new things. ZDT shared: After participating in the nature notes program, I became more careful and was able to observe and discover a lot of different things compared to other students. For example, I was able to observe the changes of the seasons, the growth of different plants during each season such as the growth of new shoots of saplings when spring arrives. I was able to observe these things carefully so that I could discover many beautiful things and feel the charm of the nature.

Develop Scientific Inquiry Skills Participating in the nature notes program not only provides students with an in-depth understanding of the traditional Chinese culture, but also expands their knowledge and encourages them to apply knowledge to their daily lives. Students are encouraged to go out into nature, talk to flowers, birds, fish, insects, and all things in nature, establish a connection with nature, and at the same time develop the ability to inquire about what is new to them.

By going into nature and observing and exploring nature, students not only enrich their scientific and cultural knowledge and deepen their understanding of knowledge, but also learn the methods of scientific inquiry and develop their scientific inquiry abilities. ZW stumbled upon a bug during an expedition to Maanxi Creek. She brought it home, observing the bug's daily growth, taking photos, writing down her thoughts, and recording the weather, date, temperature, and other information about the day. After 25 days of observation, the entire process of the worm turning into a butterfly was recorded with 130 photos. She then selected several photos that best represented the change of the worm and published them with text about her thoughts into a poster that illustrated the complete metamorphosis of the insect. ZW commented on the impacts of the nature notes program on her:

> Participating in this project has enriched my life and expanded the scope of my knowledge, such the production of multimedia, the experience of the defense process, and oral expression and communication skills. After participating in the Nature Notes program, all of these intangibles have been promoted and improved. This program has greatly helped my growth. I have gained a lot from it.

In the process of working on her nature notes projects, ZW carefully and meticulously made observations and records, and constantly found problems and solved them. She gained knowledge and practice in scientific research methods such as observing, recording data, organizing data, representing data using illustrations, and annotating. Such hands-on and mind-on projects improved students' interest and engagement in the inquiry of nature and enhanced their capacity of scientific attitude and inquiry skills.

Change the Way of Thinking and Improve the Ability of Thinking Nature offers great opportunities for students to learn biology, chemistry, geography, and physics. The scientific inquiry methods and scientific thinking learned in the nature notes projects help students optimize their ways of thinking in classroom learning. ZDT explained: Before participating in the nature notes program, when the biology teacher asked us to investigate something, I would just blindly think about it, not knowing where to start. Now, I am able to work on my biology projects through a rigorous, rational, and goal-oriented inquiry process. I can link my observation of simple subjects to the big ideas. I can reflect on what I observed and get inspiration from what I saw. Most importantly, I became more interested in discovering new things.

Students were able to apply the scientific thinking developed through the nature notes program to their learning in the classroom and gained a clearer and deeper understanding of the subject knowledge. Subject knowledge such as biology, geography, chemistry, and physics come from nature, but also move beyond nature. While working on the nature notes projects, students move from book to nature and nature to book. This is a process of connection between theory and practice, learning and inquiry, which lead to a clearer understanding of knowledge learned in the classroom and an authentic experience of scientific inquiry that classroom instruction cannot often offer. Often, the knowledge learned from participation in the nature notes program goes beyond students' classroom learning scope. This extended learning process can lay the foundation for subsequent school classroom learning and increase students' confidence and interest in school learning.

When students took nature notes, they actively learned from their own inquiry in nature instead of solely learning from a teacher-centered classroom instruction. The biggest gain was their experience with the new way of knowing. ZDT said,

> When I first joined the nature notes program, I lacked training in scientific thinking and had immature views on the things I observed and the problems I faced. For example, I found a frog with a strange sound on a field trip and told my teacher about it. My guide teacher told me that not all frogs are the same frogs as we usually see. What I saw was a tree frog, different from the frog I usually see in my daily life. From this trip, I learned that when looking at things we cannot look at them one-sidedly, we have to look at them with objective and rigorous thinking. When facing problems, we have to think about them in a rigorous way.

In the process of taking nature notes, the teacher guided the students to look at the problem more objectively and critically, to think more carefully and deeply in relation to their existing knowledge. This process enticed their thinking, pushed students to consider all aspects of the problem, and promoted their sense of exploration and innovation in nature.

Improve Students' Collaborative Learning Ability Most nature notes projects were carried out by a group of students, or several groups of students worked on the same project at the same time. ZDT recalled, "When I first joined the Nature Notes project, I was mostly independent, and I thought I would have more freedom and time to complete the project without the time constraints of everyone else." Then ZDT added, "I got more inspiration after sharing my feelings and gains with everyone. And in the process of drawing nature notes, students shared guidance with each other and gave each other some suggestions, and then added their own feelings, which not only improved the efficiency of drawing nature notes, but also improved the quality of the nature notes work." Like ZDT, participants realized that any individual person has limited knowledge and experience, while working together can lead to more knowledge, inspiration, and interest. It will also increase students' reasoning and communication skills. The positive outlook on collaboration developed from the program will benefit students in their future life and career. Develop Students' International Perspectives and Improve Their Interpersonal Communication Skills The nature notes program is a kind of open learning. It is no longer limited to school or classroom, and instead moves into society and nature. It can improve students' confidence, enrich their emotions, and improve their interpersonal skills. Student participant MYL shared, "when we were outdoors investigating wetlands or polluted rivers, we sometimes encountered local people's lack of understanding about what we were doing. We needed to communicate with them and make them understand the meaning of our activities." In such communications, MYL explained the activities to others in detail and the necessity for becoming advocates to protect the ecological environment of the river. In the nature notes program, students were able to engage with the community in a way that is not available on the school campus. Student participant ZW also mentioned in the interview:

> Through the nature notes projects, I learned how to get along with others, to communicate with others, to think differently and to take into account the feelings of others. When I made an arrangement, I first considered whether the appointment time was convenient to other people, followed by whether the location chosen was appropriate, and whether the instrument was being used by others. I learned to integrate all these factors into my arrangement, instead of just thinking myself.

In the process of working together, students learn to communicate with each other and create a harmonious atmosphere of collaboration to accomplish a task together. As a person in society, communication between people is essential, and the development of interpersonal skills is not only helpful for study, but also for life and work.

Based on the platform of the Canada-China Reciprocal Learning Program, the nature notes program exchange activities between Canadian and Chinese students does not only allow students to share different natural and biological knowledge of different countries, but also to exercise their written expression skills, enhance their confidence in communication and learning, and learn the different cultural customs of both sides and the charm of different language expressions.

Promote Students' Self-Development

Environmental Awareness and Behavior The nature notes project not only enriches students' knowledge and broadens their horizons, but also stimulates their awareness of protecting the ecological environment. ZDT, the youngest River Caption in Chongqing, and her volunteer team often carried bags with them when they patrolled the river to pick up trash. They recorded their thoughts about protecting the river in notebooks. The awareness of environmental protection was deeply engraved in their minds. These students put their love for nature and nature protection into action. HYF expressed her joy in the interview, "the excellence award I got in Chongqing nature notes contest gave me a lot of confidence and motivation to love the nature and life more, to observe the world with a more exploratory eye, and to enrich my after-school life." Through the nature notes projects, students connected their observations of nature with the 24 Solar Terms, linking education with nature and real-life experience, and developed students' independence, confidence, self-improvement, and self-management skills. The nature notes program is a way for students to learn in and about nature. Participating students developed a sense of environmental protection.

Career Planning and Professional Selection The impacts of the nature notes program on students extends from middle school and high school to college and even to their future career choices. HYF, a former program participant, is currently a senior at Sichuan University majoring in agricultural water engineering. Her participation in the program had a great impact on her major choice. She participated in the school-organized field trip to Jinfu mountain in 2014. During that trip, she observed different animals, including red-dotted toads and black-belted snakes, which triggered her interest in biology and ecology. She wanted to pursue a career in the use and protection of water resources. ZW, who fought her way through the environmental engineering program, was also deeply influenced by the nature notes program. She says, "The spirit of the nature notes program is that you have your own perceptions and feelings and then share them with others." Growing up in the countryside, her initial curiosity about nature became her internal motivation to participate in the nature notes program. Based on the current employment situation, her parents advised her to choose a popular major such as finance or economics, while she preferred to choose her favoured major in environmental engineering. In order to be admitted to her favoured major, she used a clever way to apply for a volunteer position and was eventually admitted as she wished. Driven by internal motivation, she continued to explore and research nature, discovered a new species of black-framed toad, won the Grand Prize in the nature notes competition, and received positive feedback from government departments; all of which pushed her to study and research further in the environmental field.

Life Philosophy and Survival Wisdom "Survival of the fittest" is not only the law of survival in nature, but also the law of survival in human society. WYN's observation of birds made her realize this. From nature to life, participating students not only applied the wisdom they learned from participating in the nature notes program, but also learned the philosophy of life and experienced the meaning of life. From observing puppies eating corn, to reflecting on how to care for animals in the right way; from observing the growth of tadpoles, to understanding the magic and greatness of life; from observing the process of giant phoebe butterflies changing from ugly to beautiful, to understanding that things should not just be looked at superficially.

The Impact of the Nature Notes Reciprocal Learning Program on Canadian Students

In order to examine the impacts that students received from participating in this project, eleven students participated in two face-to-face focus group interviews to share their thoughts and feelings. One focus group took place with six Grade 8 students, three females and three males. The other group had five participants, all females. In general, all 11 students felt that the 24 nature notes project was a positive experience for them. They were delighted with this unique learning opportunity. Their reasoning is summarized into three themes:

1. Participants valued the chance of experiencing nature and learning new knowledge;
2. Their personal interest and freedom of learning was fully respected;
3. The project had opened their mindsets.

The Chance of Experiencing Nature and Learning New Knowledge

There were four steps in the nature notes projects, including outdoor observation, data collection, display findings, and communication with sister school partners. Among all responses, seven out of eleven students responded that designing a way to show their learning was their favorite step among the four. They found that putting information together and building a model to present their findings was a fun experience. According to the students' sharing, the nature notes program was a new experience for them. They described some uniqueness that this program had given them. Instead of finding information online as they usually did, students went outside and observed nature. This gave them an actual chance to experience animals or insects on their own, rather than the traditional way of learning from a teacher. Through the whole process of conducting this project, students learned not only knowledge about nature but also a variety of things, including learning how to write paragraphs, how to find information, how to make a bibliography, and how to work in a team. More importantly, they knew that learning about the specific insects or animals in their surroundings would be beneficial to their studies. Student K1 learned that the common house spider "is not poisonous at all," and therefore he will not fear this insect anymore. Student A1 was satisfied with the project because "you get to learn about different animals, insects, and plants in the world around you." Student J1 also enjoyed a good learning experience in the area near her community. She explained, "we were able to connect to the things that were in our area and experience those things in the outdoors" (Fig. 5.7).

Fig. 5.7 Canadian students outdoors looking for interesting topics of inquiry

Respect Personal Interest and Freedom of Learning

A major feature of this program was that the Canadian students' personal interests were highly respected. Students valued this aspect, as it helped them become self-motivated learners. Besides this, the students were extremely satisfied that they were given complete freedom in selecting topics, as well as designing a way to present their learnings. They indicated that they "had never experienced such full freedom in other assignments." Students particularly liked to have a very broad topic to choose from. They were encouraged to pick whatever topic that interested them. Once they picked a topic, they started their learning observing nature and researching more about the selected topic. After that, students can narrow the topic down and collect data or information. The last step for them was to design a way to present their findings. Student A1 specifically indicated that the teacher allowed them to pick their own inquiry question. "Unlike the other projects, our teacher picks inquiry question for students," she said. Student H1 also shared that she was allowed to pick her own animal or insect, unlike other projects where all students were required to research the most common animal. H1 preferred this way because those teacher-centered projects were "quite boring." Student I1 indicated that the other projects were "structured," and students had to write "long reports." Student J1 also liked that the research topic was not assigned by the teacher but chosen by the student. She enjoyed the diverse options that she could choose from. In this project, the student was the initiator of their own learning. The teacher respected each student's personal interest and set them free to do what they felt was interesting. These two points were essential as they kept students highly motivated. Consequently, students were willing to complete their projects in a timely manner. Take student B1's response as an example, she said that she could do it any way she liked to present her work. All Canadian students highly appreciated the freedom, which helped open their imaginations and create what they felt was appropriate to showcase their learning. To summarize all Canadian students' sharing, with the unique way of doing this program, students had learned how to pick inquiry question(s), how to gather information, and how to answer those questions. Those achievements fulfilled the curriculum criteria, as well as met the program requirements. Moreover, the inquiry topic that they chose reflected their different personal interests. The creativity shown in their final products highlights the benefits of the student-driven process of learning. Four pieces of student work are provided below as examples (Figs. 5.8, 5.9, 5.10, and 5.11).

Fig. 5.8 Showcase 1: Student's inquiry in bees (Inquiry question: How do honeybees survive in winter?)

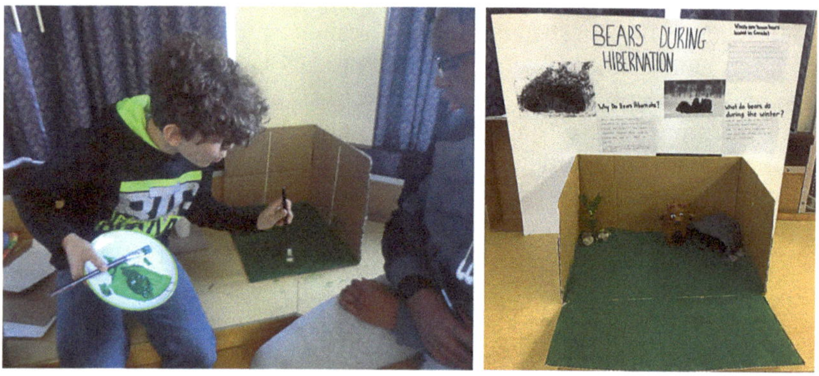

Fig. 5.9 Showcase 2: Student inquiry in the hibernating bear (Inquiry question: Why do bears hibernate in winter?)

Open Students' Mindsets

Canadian students shared that the nature notes program had opened their mindsets and given them the opportunities to learn from their counterparts in China. The students went to the school garden to look for animals or insects that interested them. This was their initial way of connecting with nature. Unlike their Chinese partners whose parents normally provided transportation to conduct nature observations, these students' parents were not always able to bring them to a more natural environment

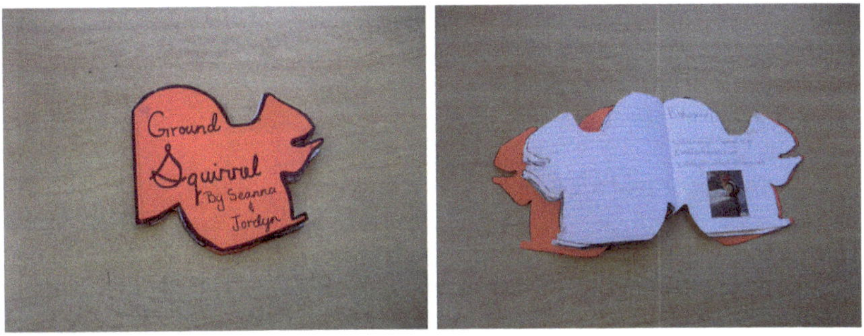

Fig. 5.10 Showcase 3: Student inquiry in squirrels (Inquiry question: When and how ground squirrels decide to go for hibernation?)

Fig. 5.11 Showcase 4: Student inquiry in butterflies (Inquiry question: Why are butterflies an important part of nature?)

due to work or other reasons. Instead, the Canadian students had to find their own way to connect themselves with nature. For instance, student K1 was interested in spiders, but he first thought that the common house spider was poisonous so he avoided touching them. However, K1 decided to do a science project with this insect because he believed that "it's a good project to open up your mindset." He first did some research into the common house spiders and found that they are actually not poisonous. This discovery gave him confidence to catch a spider from his basement and see how long it would live. Student K1 spent a period observing the spider and also feeding it with different foods. He later had some findings about the food that the spider eats. K1 documented the process and planned to show his pet spider and all his findings to the Chinese partners in the Skype video conference. Unfortunately, this pet spider died for unknown reasons. Student K1 was very upset about that, however, he learned a lot of new knowledge from doing this project and his mindset had shifted.

According to the two Canadian teachers, their students loved the idea of "life lessons" which was shown in the Chinese students' project work. In a Grade 8 class, teacher H had a discussion with the students after having the video conference, one of her students told her that the Chinese students "included a moral, like the legend building" in the project. The students hoped teacher H could incorporate that idea in the next round of their projects. Teacher F's students also had such hopes. In response, teacher F added two new questions which helped her students looking at the animals, plants, or nature from a new perspective. Her students welcomed this addition. In the group interview, all Grade 8 students reported that they liked the idea of making stories to reflect on what they learned from nature. Student D1 shared her view by giving an example. She commented that the Chinese students "always have really interesting stories. Like that one girl who talked about the tree [the tree in her backyard was struck by lightening. She was very sorry for the tree. But the tree surprisingly came alive in the spring. The student commented that a person should not give up]. That was like a real story, and it was cool." Student A1 elaborated her feelings, "it is easier for you to relate to the animals when you learn the similarity between you and the animals." Take another story as an example, a Chinese student used to have a pair of pet parrots. One day, one parrot escaped from the cage, but it did not fly away. The Chinese student interpreted this as the parrots staying loyal to each other because the one that escaped was waiting for the other one to be set free. Canadian students appreciated this story, but they had a little bit different interpretation. Student B1's reflection was, "you should stay loyal to someone accompanying you. You have someone in such a lower situation, you have to still stay with them and help them." This example showed that forming a life lesson like their Chinese counterparts was a brand-new experience to the Canadian students. Canadian students appreciated the Chinse students' inspiration in creating stories related to human life. They were also willing to try the Chinese way of perceiving life from a different angle. For instance, Canadian student J1 studied caterpillars (Fig. 5.12), and she shared a life lesson that she learned, "people should not judge other people based on their appearance." This example demonstrated that

Fig. 5.12 Caterpillars

the student's understanding gained from this activity stepped forward from just simply doing a regular project, to forming a deeper understanding of life. This change suggested that Chinese students' work had a great impact on how Canadian students viewed the human-nature relationship.

Cultural Differences and Mutual Impacts

In this program, Canadian and Chinese students realized the differences between them. Chinese students often expressed their reflection on their nature observation by connecting nature with their own lives or human society in general. Canadian teachers and students were inspired by this new aspect of outdoor learning. For example, Canadian student J1 took a life lesson from her observation of how "ugly" caterpillars turned into beautiful butterflies.

Chinese students realized the strength of Canadian students as well. Canadian students demonstrated great creativity in presenting their works. For instance, their innovative ways of presenting can be illustrated by a 3-dimensional butterfly (Fig. 5.13), a ladybug-shaped cake (Fig. 5.14), and a clay model hedgehog in a shoebox (Fig. 5.15).

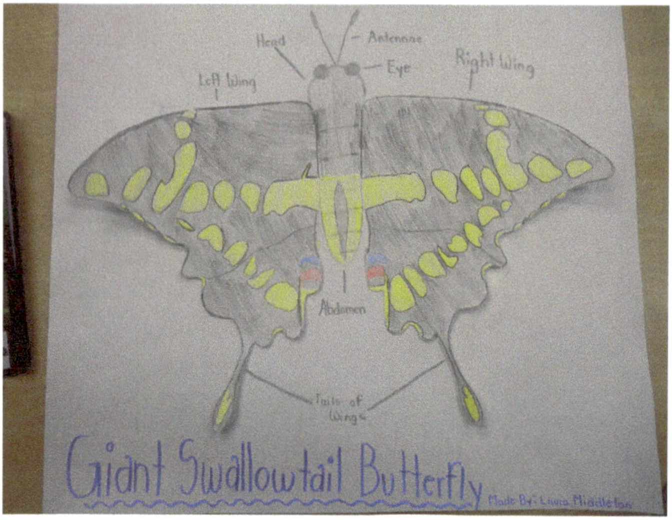

Fig. 5.13 Giant Butterfly Three-dimensional Painting

Fig. 5.14 Ladybug Cake

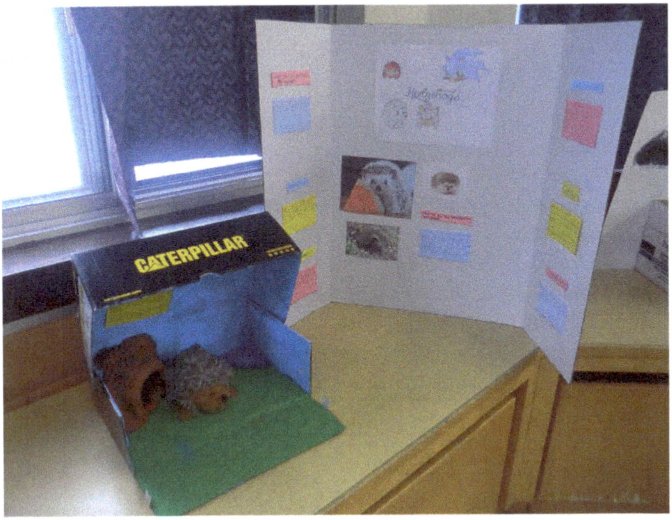

Fig. 5.15 Hedgehog in a Shoebox

During the Skype meetings, Canadian students noticed the differences between themselves and the Chinese students in terms of presenting what they learned. They appreciated the Chinese students' talent in art, as they always drew beautiful pictures. However, student D1 proposed a critical point, "they all presented their work in the same way. They all drew pictures." Unlike Chinese partners, Canadian students presented their work in many different ways. Canadian students were proud of their own creativity because it "makes the project more fun." Since they noticed these differences, Canadian students gave some constructive suggestions to their Chinese partners. They suggested Chinese students "change the way of presentation, not just doing it like a picture and writing." The Canadian students hope to see the Chinese students' uniqueness in their ways of presentation because they wanted to help them become creative members of the learning community.

Reflecting upon each other's strengths and weaknesses made the program a valuable reciprocal learning opportunity for students from both sides. Since its inception, the nature notes program has not only improved students' knowledge and abilities but has also demonstrated its unique characteristics in terms of self-development. The nature notes program has been a platform for students from both sister schools to exchange and learn about their respective cultures, especially in science and technology.

References

N.d. (2019, October 23). *Chongqing's youngest river chief: A 13-year-old who found new species was greeted by an UN official.* Chongqing Evening News Website. Retrieved from https://baijiahao.baidu.com/s?id=1648182902715068531&wfr=spider&for=pc

Tao, X. Z. (1934/2016). *Actions are the beginning of knowledge.* Gu Wu Xuan Publishing House

Tang, S. C. (2013). The specialized curriculum and professional orientations for high school students. *Educational Development Research, 18*, 15–21.

CHAPTER 6

Teachers' Experience and Perceptions of the Program

Sonia Ho, Guihua Zhang, Yanhong Mao, Han Yang, Xin Chen, George Zhou, and Yuanrong Li

To collect participating teachers' experiences and perceptions of the project, the Canadian and Chinese research teams conducted interviews with the participating teachers (Table 6.1). On the Canadian side, interviews were conducted at the end of the second year, and the middle of the third year of the program. Two Canadian female teachers participated in one-on-one semi-structured interviews. One taught a grade 7 and 8 split class, and the other teacher taught a grade 8 class. Both of them were not science-specialized teachers. The Canadian research team also interviewed two Chinese male science teachers (L and X) individually through

S. Ho • G. Zhou (✉)
University of Windsor, Windsor, Canada
e-mail: gzhou@uwindsor.ca

G. Zhang • Y. Mao • H. Yang • X. Chen • Y. Li
Southwest University, Chongqing, China
e-mail: hxliyr@swu.edu.cn

G. Zhou et al. (eds.), *Science Education and International Cross-Cultural Reciprocal Learning*, Intercultural Reciprocal Learning in Chinese and Western Education,
https://doi.org/10.1007/978-3-031-17157-4_6

Table 6.1 Teachers who participated in interviews

Teacher	*Country*	*Gender*	*School subject*
F	Canada	Female	Generalist
H	Canada	Female	Generalist
L	China	Male	Biology
X	China	Male	Geography
W	China	Female	Language art
S	China	Female	Chemistry
Y	China	Female	Chemistry
T	China	Male	Biology

teleconferencing. On the Chinese side, four teachers were interviewed. Two in-person interviews were conducted with two female teachers (W and S). Two survey interviews were conducted with one female teacher Y and one male teacher T. Teacher W was a language teacher and an early participant in the nature notes program, S was a chemistry teacher, T was a biology teacher, and Y was a chemistry teacher. These interviews captured teachers' perceptions and experiences toward various aspects of the program. Teachers recalled their experiences of preparing and conducting the program, students' performances, change in perception towards each other's culture, their learning from sister schoolteachers, and their concluding thoughts about international collaboration. The gathered interview data provided sound information to explore the following research questions:

1. How did participating teachers experience the program?
2. How did the program impact them?

Teachers' Perception of the Program Implementation

Both Canadian and Chinese participating teachers highly valued their experience with the program and reported that the program had a significant impact on their students and themselves. On the Chinese side, the project had been ongoing for two years before it was officially introduced to the Canadian sister school. According to Teacher L, the program was first launched in their school and later became a popular program in the local city. It was designed to be implemented in two dimensions: an integral component of the mandatory science curriculum or a specific topic for elective courses. Students who participated in the program conducted their projects during school hours or in their spare time. The roles that the Chinese teachers took in the program included taking students to field observations, helping them pick up an interesting topic, and facilitating them to study their chosen topics. The school held competitions to motivate them to participate in this program.

At the beginning of the academic year, the school introduced the nature notes program to all new students so that every student at the school was aware of the program. The first-year high school students were required to take a course that orientated them to learn through research. This course lasted seven weeks. During this course, the teacher introduced the nature notes program to students. According to Teacher W, students were very interested in the program. Approximately 60–70% of new students would contact their teachers after the course to inquire about the program. Interested students were assigned to a guiding teacher. There were over 60 guiding teachers in the school to facilitate students' nature notes projects. Students worked on their projects throughout the school year, including the winter and summer holidays. The following September, students submitted their finished projects for the school competition. Top works would be recommended for competitions held at the city and national levels.

Participating in the program was rewarding to both the teachers and students. Chinese teachers made every possible step to make the program a successful experience for their students. Chinese teacher L explained that "we reward them if students show excellent work in this program… Their greatest projects helped some students to get admission offers from key universities." This is especially attractive to senior students as the program opens a better future for their higher education. Students became confident learners through their participation in the program as they received recognition from the school and teachers. The guiding teachers often felt rewarded when they observed the progress students made in their projects and heard the news of student projects being recognized in the competitions.

Teacher X indicated that this program was designed as "a remedial for the shortage in traditional education," which "fostered students to observe more carefully." Teacher L felt that students who participated in the project showed higher interest in science than those who did not get involved. In short, students' self-motivation toward science learning increased.

The nature notes program expected students to gain a deeper understanding of the real world through a combination of book knowledge and nature observations. Teacher W described,

> I started the program in the third year, and the overall feeling is that the number of nature notes projects has increased because of the publicity. In the beginning, the students did not know much about the 24 Solar Terms. When I took over, none of the first-year students knew much about the 24 solar terms. I wondered if I could improve the students' knowledge about them in my language classroom. In conjunction with the nature notes program, I asked the students to draw the 24 solar terms. The tasks were assigned for the months of July, August, and September. As far as I know, the integrated biology and geography practice class advocates observation at any time, so the observation of the nature is extended to all seasons of the year. For different subject teachers, the focus of the nature notes program is different. For example, geography teachers are more about geography. Biology teachers are more about physical characteristics, the development and growth of the characteristics of plants and animal. Language teachers implement the program in the context of language education.

Teacher T had a similar feeling, "before participating in the program, students knew little about the culture around the 24 Solar Terms. After participating in the program, they showed a strong interest in traditional culture and were happy to share what they had learned with their parents." Teacher W said that this project had different effects on students of different grades. For the lower graders, this project inspired their interests. Student observation skills were improved while they got close to nature. Their understanding of nature was bettered. Since there was no competition for the lower grades, the program focused on developing students' abilities. In contrast, the upper graders have a more explicit purpose for participating in the project, hoping to participate in the nature notes competitions and win awards. Therefore, the senior students gained more experience and more profound knowledge from participating in the program. Their sense of inquiry was stronger, and their research was more in depth than junior students. Teacher S reported that all students who participated in the natural notes program improved their abilities even though the level of improvement was different among students. Students with greater academic records were able to balance their time between their subject studies and nature note projects, and could complete both their studies and projects more efficiently. Students with poor academic records faced challenges in allocating their time between their studies and projects. Some struggled to complete project requirements in time even though they were motivated to participate. Teacher T also believed that all students who participated in the program benefited. They all learned new knowledge, accumulated experience, and gained growth. This teacher pointed out students' learning about the 24 solar terms. He noticed that students did not know much about this traditional culture before. After participating in the program, students became more interested in traditional cultural knowledge and were willing to learn it and study it. In addition, students were eager to share the knowledge they learned with their families, relatives, and friends.

Beyond the academic impacts, the program profoundly influenced other aspects of student life. Teacher X shared one student's amazing experience when she and her dad went out in nature for observation:

> A student usually had little interaction with her dad, who was a businessman. One day, her father took her to observe nature and shared some childhood stories with her. She found that her dad was a very interesting guy, which made her feel that her relationship with her father had changed. After that conversation, she had more interaction with her dad.

Teacher X commented that this project helped foster parent-child communication and improved their relationship. He felt touched as "this important achievement is beyond the educational goal that was originally set for the program." Teacher W pointed out a unique session of the nature notes contest, which encouraged parents and their children to finish projects collaboratively. Parents and children need to observe the chosen topic together and integrate their findings into one visual and textual presentation. It is common sense that parents and students often have different focuses of observations. Children are more focused on funny, intuitive, and visible aspects of nature, and parents are more interested in plant growth and environmental protection. It was the collaboration, negotiation, and creation that brought children and parents closer.

The ways Canadian students presented their findings impacted the Chinese students. Chinese teachers reported that their students became more creative and innovative in presenting their work. Beyond adapting the formats of Canadian students' presentations, including 3D models and posters, Chinese students took a further step to incorporate Chinese traditional skills, such as papercutting art, in their project presentation.

Through the reciprocal learning project, Chinese students improved their skills in observations, demonstrated in-depth thinking, and fostered personal reflections. Teacher L shared that "Through nature observations, they sincerely considered the relationship between human and nature. I found that most of these feelings (towards human-nature relationship) were formed naturally." Their interest in nature not only increased, but also continued through their high school career. Teacher L shared that "After participating in the nature notes program, students paid more attention towards the things that surrounded them. Even though some of them were not interested in participating in competition, they still did it for their personal interest." The students who had participated in the exchange meetings and shared their projects were very proud of themselves. They appreciated the opportunity to share their work with their international counterparts. They saw their capability and potential to share research in a cross-cultural context.

In the Canadian school, teachers integrated the 24 Nature Notes project into their literacy curriculum and made it into an inquiry-based learning project. They combined both teaching goals into one single project, and students had to fulfill successful criteria that covered aspects from both curriculums. Students were allowed to choose to either work in groups or independently. The grade 7/8 split class teacher designed a work package to help students sort out their thoughts and ideas and prioritize their tasks. The Grade 8 class also used this work package for their project. The teachers later found that it helped increase students' effectiveness. Whenever students needed guidance, they could refer to the practice sheets included in the package or ask their teachers. The teachers provided instructions according to students' interests and their research plans. Students were allowed to take the package home during breaks to continually observe nature, collect/record data, generate ideas, and work on their presentation during the break.

Regarding the impacts that the nature note program had on Canadian students, Teacher H indicated that the program increased her students' engagement level and got them to express themselves creatively. "It was great exposure to new ideas and things they haven't got to experience … without the program." They worked very hard and enjoyed the program; overall, it became a positive learning experience for the students.

Teacher F shared a similar perception with teacher H. Her students not only learned to do an inquiry project, but they also learned a lot about the research process itself. After seeing how the students and teachers were doing this program at the other end of the world, Canadian students learned to think in a different context and made reflections in a different way. This achievement is beyond just the academic sense. "The exchange of understanding on how different cultures work … is so beneficial" to her students, Teacher F commented, "we are still isolated in the understanding of how people work outside of our own culture. The ethnocentric type of viewpoint becomes really dominated." The nature notes reciprocal learning program opened her students to "actually have exchanges of dialogues with other people who are just like them but doing things completely different" while it is still working. Stepping out of that ethnocentric viewpoint is the most impact that her students got out of the program.

Canadian students presented their work periodically to their counterparts during video conferences. The creativity of their products, the confident attitudes that were shown in the video conferences, and the satisfaction that was demonstrated in the interviews together lead to the conclusion that the program was successfully implemented by the teachers in the Canadian school.

Teachers' Professional Growth

Participating teachers from the two sister schools highly valued the project as a positive learning experience. Canadian teachers enjoyed participating in this program because they learned to view nature from a new angle. Teacher F liked the whole idea behind the nature notes program associated with Chinese culture. She enjoyed communicating with the Chinese sister school. She felt that it was very interesting to experience the same thing as the Chinese teachers did but in entirely different ways. She gave a detailed explanation to her thought:

> This program really forced me to look at things in different ways that I haven't done in the past … Every time that we would share with our Skype meetings, and I would learn something new … I was able to have discussions afterwards about the things that we have learned … I think that kind of self-reflection aspect is very valuable as a teacher.

Teacher F liked this program because she expanded her own knowledge through her participation. To Teacher H, this was a program project that she "would not have thought about doing" in her class, and that she "was excited to try something new." She viewed the program as a great learning experience, not only for herself but also for her students. She reported that the experience of participating in this program was beyond her expectations.

For Chinese teachers, the positive impact they gained was to have "actual practice" in the science field. Teacher L reported that the program allowed him to "pay attention to the nature and to the creatures" that they had seldom noticed before. Teacher T said that this project has taught him about traditional culture, how to guide students to make effective and interesting nature observations, and how to develop students' ability to make discoveries. Teacher W said that this program motivated them to learn new things because he needed to have a wide range of knowledge to guide students in the program. Teacher S was a laboratory teacher who taught a research-based classroom and infused his laboratory teaching with the idea of observing and experiencing nature. She also said that she had to learn more in order to guide her students' nature notes projects because students were often interested in topics that she was not aware of before. She believed that as a teacher, she needs to broaden her knowledge. She learned multiple subjects and disciplines on her own, and she grew with her students as they participated in the nature notes program. Teacher Y said she learned more about traditional Chinese science and technology culture through this program. She said she had no idea what solar terms were before joining the program and probably could not even name a few of the 24 Solar Terms. However, after participating in the program, she gained a better understanding of the names of the 24 Solar Terms, and the weather, customs, and culture that these names represent. She shared her understanding of the nature notes program as a project-based learning experience that required students to observe, record, and reflect on nature. Participating students learned through firsthand inquiry experience with nature.

In addition to their professional improvement, Chinese teachers received recognition through the nature notes program. For example, in the First National Youth Nature Notes Competition in 2018, their school won the excellent organization award, and six teachers won the mentorship award. In the Second National Youth Nature Notes Competition in 2019, the school won the excellent organization award again, and seven teachers won the mentorship award.

Respecting and Learning from Each Other's Culture and Practice

The partnership program ran for three school years, from September 2015 to May 2018. In the first year, the Canadian school spent a year exploring some of the possible ways to integrate the project into its curriculum. A breakthrough came in the 2016–2017 school year. A meeting was set in November 2016 between the research team and two Canadian teachers who were willing to participate in the project. They received background information about the project. While they individually learned the background knowledge, they consulted the research team whenever they had questions regarding their plans for the implementation of the program in their classes. The teacher participants appreciated the chance to learn about Chinese culture. Teacher F reflected:

> I felt that I wasn't aware of how the season changes in China before this. I have no idea about the whole aspect here, like the solar terms...It is completely different from everything I have ever learned, but that's what I enjoyed.

Teacher F believed that "self-reflection is very valuable as a teacher." She realized that a lot of things made sense after reading materials that were provided to her by the research team. "Like the Awakening of Insects … that make sense … from a cross-cultural point of view. It is very interesting for myself." She enjoyed learning about "how different cultures experience the same things but in completely different ways" because that increased her knowledge.

From the point when Teacher F took over the program, she attended every video conference and was prepared to learn new knowledge. A Chinese student shared a story about a pair of parrots that she observed in one of the video conferences. In her story, one parrot escaped from the cage, but it did not fly away. It seemed that it was unwilling to leave but waited for its partner. It flew away with its partner once the student let go of the other parrot inside the cage. The Chinese student interpreted this as a romantic story—the two parrots stayed loyal to each other. She then connected such observation with human family life and moved on to criticize the ever-increasing divorce rate in China. Another Chinese student shared her observation of a tree in his grandmother's backyard. The treetop was broken off by a storm one year. She felt sad about the tree and believed it would not survive. However, the tree grew new leaves in the following spring. She commented that a person's life could be like this. Everyone might have times of difficulty, but he or she should not give up and always stay optimistic about future opportunities. Teacher F was enlightened by the idea that the Chinese students learned such life lessons from their nature observations. Teacher F shared her excitement in that meeting: "what a fantastic idea it was." She delivered her appreciation to the Chinese teachers and students.

Teacher F took notes about the idea and decided to incorporate this life lesson into her next round of the program. As a result, Teacher F revised the student inquiry work package by adding the following two questions for the next chosen solar term, Awakening of Insect: (1). What lesson did you learn about life through understanding your chosen insect or animal? (2). What do you see differently now that you know about your insect or animal?

Through this exchange of ideas, Teacher F acknowledged the cultural differences between the two countries. She not only valued the differences but was also willing to open herself up to accept new ideas. Revising the work package, she designed and added questions to induce students to think deeper, which showed her appreciation of the Chinese idea. The other Canadian teacher, Teacher H, said that she also liked the idea of gaining personal reflection from the program. Teacher H particularly liked the life lesson idea because "kids like story and it increases their level of engagement." She "was so excited" to learn about this idea, although she did not build that into her next program design. In her future teaching, she promised to "take more opportunities to build that in."

The Chinese teachers had noticed some advantages shown on the Canadian side. The nature notes reciprocal learning program was designed as an inquiry-based program, which means that the Canadian students started their projects by asking an inquiry question, then students go to collect data, which includes outdoor nature observations and conducting online research. After they accumulate sufficient knowledge and information from the student-led learning, they can decide and present their work in a way that they like. During the whole process of the inquiry-based study, the Canadian teachers gave their students freedom and space to explore what interested them, where they could get information from, and which is the best way to demonstrate their learning. As a result, the students produced various work, including models, booklets, poster boards, PowerPoint presentations, drawings, and written work.

The Chinese teachers also highly valued the Canadian students' creativity in their demonstrated work. During one of the video conferences, a Canadian student presented a three-dimensional butterfly model, which surprised all Chinese video conference attendees. The Chinese teachers appreciated the innovations shown in the Canadian students' work. They were inspired by the Canadian students and decided to add new requirements in the next round of the program. The Chinese teachers encouraged their students to learn from the Canadian partners and explore new ways to present their work. According to the lead science teacher L, this encouragement worked very well, as they saw significant changes in their students' work. As a result, Chinese teachers collected many more three-dimensional projects than in the previous round of the project. These changes are reflected not only in the way the work was presented, but also in the number of products submitted. In other words, the quality and quantity of Chinese students' work have increased significantly.

Summary

Gillespie and Melber (2014) reported that, in an international collaborative program with a focus on science learning, the understanding of each other's differences equipped participating teachers and students with an open-minded attitude so that they could not only sense each other's strengths but also understand each other's limitations. Such open-mindedness is necessary for any participant to have in a reciprocal learning program.

Overall, participating in the nature notes reciprocal learning program was a positive experience for the teachers and students from both sides. It led to profound impacts on both the Chinese and Canadian students, as well as the teachers. The Chinese students became self-motivated learners, and their creativity and passion for nature increased. Participating in the project enlightened them to choose science study as an academic and career future in the long run. For the Canadian students, they developed a good understanding of a completely different culture. Participating in the project enabled them to see how things were operated on the other side of the world and understand other people's thoughts and feelings from a different perspective. They got this opportunity to think out of the box and had new reflections through interacting with their Chinese counterparts. Our findings are consistent with past studies revealing that students from different countries obtained successful educational experiences through international collaboration programs (Ligorio & Van der Meijden, 2007; Wang, 2012; Yang et al., 2014).

The Chinese teachers enjoyed leading the students to work in the science field and conduct nature observations. The teachers enjoyed answering their students' questions and giving them concrete guidance. Having the chance to lead the students into the research world and seeing them complete their research eventually gave the Chinese teachers a sense of satisfaction. The Chinese teachers also treasured the opportunity to work with their Canadian counterparts. Though the project sharing required translations due to the language barrier, the Chinese teachers were still satisfied with the collaboration. They saw how creative and innovative the Canadian students could be, and they also appreciated how well the program was implemented on the Canadian side.

The Canadian teachers had professional growth from participating in the program. They enjoyed learning how the Chinese people view nature from a different angle. They also encouraged their children to learn from the Chinese students because they wanted them to experience something they had never known before. The teachers also hoped their students would learn to view nature from a different perspective. They particularly liked the idea of life lessons. Therefore, they built this idea into their project requirements so that their students could generate personal reflections on life. The two teachers indicated that this impact would extend to their future teaching.

With an increasing diversity of student population in Canada, scholars have advocated Culturally Responsible Pedagogy (CRP) (Harrison & Skrebneva, 2020; Sleeter, 2011). Doucet (2017) argues that all children "are deserving of classrooms in which their humanity is seen and honoured, and in which their cultures, languages, and family histories can be bolstered and sustained" (p. 200). Gay (2010) believes that teachers' understanding of students' cultures is fundamental to school achievements and advocates a need to connect teaching and learning to the social and cultural experiences of students through CRP. Alim and Paris (2017) argue that CPR offers a means of sustaining cultural ways of being and must account for the cultural beliefs and values of students. The cross-culture learning experience and achievement through the nature note reciprocal learning program are significant for Canadian teachers. It offers a model for teachers to develop their competence in CRP as well as a model for CRP practice.

References

Alim, H. S., & Paris, D. (2017). What is culturally sustaining pedagogy and why does it matter? In D. Paris & H. S. Alim (Eds.), *Culturally sustaining pedagogies: Teaching and learning for justice in a changing world* (pp. 1–21). Teachers College Press.

Doucet, F. (2017). What does a culturally sustaining learning climate look like? *Theory Into Practice, 56*, 195–204.

Gay, G. (2010). *Culturally responsive teaching: Theory, research, and practice* (Multicultural education series) (p. 2e). Teachers College Press.

Gillespie, K. L., & Melber, L. M. (2014). Connecting students around the world through a collaborative museum education program. *Journal of Museum Education, 39*(1), 108–120.

Harrison, N., & Skrebneva, I. (2020). Country as pedagogical: Enacting an Australian foundation for culturally responsive pedagogy. *Journal of Curriculum Studies, 52*(1), 15–26. https://doi.org/10.1080/00220272.2019.1641843

Ligorio, M. B., & Van der Meijden, H. (2007). Teacher guidelines for cross-national virtual communities in primary education. *Journal of Computer Assisted Learning, 24*(1), 11–25.

Sleeter, C. E. (2011). An agenda to strengthen culturally responsive pedagogy. *English Teaching: Practice and Critique, 10*(2), 7–23.

Wang, C. M. (2012). Using Facebook for cross-cultural collaboration: The experience of students from Taiwan. *Educational Media International, 49*(1), 63–76.

Yang, J., Yu, H., Chen, S. J., & Huang, R. (2014). Strategies for smooth and effective cross-cultural online collaborative learning. *Journal of Educational Technology & Society, 17*(3), 208–221.

CHAPTER 7

Conclusion and Suggestion

George Zhou, Shijing Guo, Banglong Jiang, and Sonia Ho

The nature notes reciprocal learning program originated from a school-based nature notes curriculum initiated at the Chinese sister school, Southwest University Affiliated Secondary School. The school-based curriculum emphasized experiential learning through nature exploration. After this curriculum had been successfully implemented for a couple of years in the Chinese school, the Prince Edward Elementary School in Canada partnered with the school for the purpose of reciprocal learning. The two sister schools conducted the nature notes reciprocal learning program. The program ran for three school years, from Sept. 2015 to May 2018. Instead of designing as a researcher-driven top-down program, the nature notes reciprocal learning program was actually a school student-driven program. Through this program, students from both countries

G. Zhou (✉) • S. Ho
University of Windsor, Windsor, Canada
e-mail: gzhou@uwindsor.ca

S. Guo • B. Jiang
Southwest University, Chongqing, China

G. Zhou et al. (eds.), *Science Education and International Cross-Cultural Reciprocal Learning*, Intercultural Reciprocal Learning in Chinese and Western Education,
https://doi.org/10.1007/978-3-031-17157-4_7

picked up interesting topics from nature observations and finished inquiry-based projects. Students also had chances to share what they achieved through virtual exchange meetings.

Program Impact on Students

The nature notes reciprocal learning program had significant impacts on Canadian students. Understanding how people from China are working and thinking is a great way to take the Canadian students out of the dominated ethnocentric viewpoint. This was the greatest benefit the students had through participating in the program. Canadian students were impressed by the dedication of Chinese students to their nature notes projects and, in turn, were motivated to work on their own projects. They learned from Chinese students how to reflect on what they observed and build nature observation with their own life.

Likewise, the nature notes program had a multifaceted impact on Chinese students. First, participating in the nature notes program has enriched Chinese students' comprehensive knowledge and understanding of traditional Chinese science and technology through an interdisciplinary perspective, the emergence and development of the Chinese farming civilization, the scientific reasoning behind the 24 solar terms, and their significance to a productive life. Chinese students applied what they learned in the classroom to observe and investigate nature while acquiring new knowledge, methods, and skills in the process of observation and investigation. Secondly, during the program, Chinese students' overall abilities are further enhanced, including observing nature, thinking scientifically, investigating scientifically, and learning to think, analyze, and solve problems through an interdisciplinary perspective. Students also learn the problem-solving skills of the Canadian students and the methods and skills to observe and investigate nature and improve their interpersonal and language communication skills. Particularly, Chinese students learn the fresh ways Canadian students used to present their findings.

The nature notes program not only enriches students' knowledge and improves their overall ability, but more importantly, it promotes students' self-development and future career planning. The program effectively cultivates students' environmental awareness and behaviors. In completing their nature notes works, students experience the magic and greatness of life and realize all lives have a sense of feeling, which leads to awareness of protecting the natural environment and building the harmony between humans and nature. Through their participation in the project, some students have thought more clearly about their choice of major and career aspiration. Finally, through the nature notes reciprocal learning program, Chinese students learned some new knowledge about Canada that their textbooks did not teach them.

Program Impact on Teachers

The nature notes reciprocal learning program provided Canadian teachers an opportunity to learn something that was completely different from their original understandings. Knowing how the season changes from the Chinese perspective increases their knowledge of Chinese culture. Canadian teachers enjoyed learning new knowledge and enjoyed understanding people's thoughts and reflections from a cross-cultural perspective. They treasured this experience because it was interesting to them to see things from a new and different viewpoint. This cross-cultural experience enriched their teaching, and they likely made nature notes as part of their regular teaching plan. While they had their unique ways of implementing nature notes in their classroom teaching, the way Chinese teachers did them offered some new insights. Like their students, Canadian teachers were impressed by the life lessons Chinese students and teachers took from their nature observations.

The program has positively impacted Chinese teachers in many ways. The Chinese teachers who participated in the program included both in-service and pre-service teachers. The in-service teachers were mainly teachers from the sister school participating in the program, and they came from the disciplines of biology, chemistry, geography, language arts, etc. Through the program, Chinese teachers formed an interdisciplinary learning and research community, which expanded the interdisciplinary vision of the subject teachers and their ability to develop curriculum resources in an integrated manner. Through preparing, organizing, and implementing the curriculum, Chinese teachers promoted their ability to design and implement an integrated curriculum. In this process, they needed to think about how to guide students to carry out effective and engaging research-based learning activities, develop and utilize curriculum resources of related subjects, and integrate the natural observations with the subject curriculum. Chinese teachers appreciated the results achieved by the Canadian school, teachers, and students through the implementation of the program. The exchanges and interactions between China and

Canada have facilitated a change in teachers' teaching philosophy, actively trying to shift to an interdisciplinary, integrated, and inquiry-based approach to teaching. In this process of collaboration and communication, not only did they learn more systematically about their own country's traditional science and technology, but they also gained an understanding of Canadian society and school education and deepened their understanding of the importance of developing international cooperation.

The participation of graduate students (pre-service teachers) in the program is part of the philosophy of the more extensive program: learning to learn, research, and grow professionally while doing the program. The pre-service teachers involved in the project consisted mainly of graduate students from the Chinese science team. For the pre-service teachers, the program had a positive impact on them in many ways. During the interaction, the Chinese pre-service teachers found it challenging to communicate with each other in language. Faced with the challenges, they motivated themselves to enhance their English language learning, especially their listening and speaking skills, to achieve better communication. In the process, their English communication skills improved significantly. Through participating in the program implementation at the Chinese sister school, they learned how to use the natural environment and turn it into a curriculum resource and how to design, organize and implement effective and engaging research-based learning activities. They also improved their ability to conduct research, enhanced their awareness of research ethics, and broadened their academic horizons. In addition, the Chinese pre-service teachers deepened their understanding of traditional Chinese science and technology culture through the program. They gained an in-depth understanding of Canada's multicultural background, educational philosophy, and teaching methods, broadening their international perspective in the field of science education.

Student Suggestions for Better Collaboration and Practice

Considering the format of work presentation, Chinese students suggested using computer technology to help with presentations. They suggested an online format to document nature notes, making it easier to store works for a longer period of time while saving paper. When collaborating with the Canadian schools, Chinese students suggested using new technology to facilitate communication and presentation. They hoped that using technology would break through the time limit. Chinese students appreciated the creativity and passion that the Canadian students demonstrated for the project. They suggested that Canadian students do more nature observation and collect first-hand data. They also suggested promoting one-on-one interaction between Canadian and Chinese students as it would help them to focus on a specific topic that they like in common. Creating this type of relationship will make the communication more accessible and more frequent. While they greatly appreciate the guidance they received from their teachers and the support from their parents, they hoped that their parents would not be overly involved in their projects. Finally, Chinese students hoped that more Chinese students would be able to participate in the Nature Notes reciprocal learning program in the future so that more students could benefit from the program.

Canadian students suggested the future program covers a larger variety of subjects, such as the interaction between plants and animals and between human beings and nature. They wanted to go to different places for observation instead of limiting their observation to the local community and school yard. The broader area of observation could help them get more involved in the project and collect first-hand information about a variety of nature phenomena. In terms of the method of presentation, Canadian students enjoyed the freedom to choose how to present their work. They suggested that their Chinese counterparts diversify their ways of presentation. Canadian students noticed that Chinese students made good connections between nature observation and their life and often proposed an inspirational reflection on what they observed. Like Chinese students, Canadian students would like to see a pen-pal type of relationship with individual Chinese students so that the pair of students from both sides could exchange their work and ask each other questions throughout the process. Canadian students also hope more students would get involved in the cool exchange learning experience. They enjoyed the project and wished students from different grades would have this learning opportunity. They hoped to have a longer period of time to share the projects in the future.

Teacher Suggestions for Better Collaboration and Practice

The Chinese teachers acknowledged several limits in sharing students' work caused by the time difference and the quality of the network and equipment. They provided some suggestions hoping to tackle these issues in the future implementations of the program. They suggested setting up a social platform where students could store their work so that both sides can access each other's work at a convenient time. Student projects could be sent to each side prior to having the actual Skype meeting. In doing so, both sides could have sufficient time to look closely at each other's work and then come up with questions, which would shorten wait times during the meeting and increase the quality of communication. Students will be able to generate more ideas through discussions. In addition to students' communication, teachers' communication can be strengthened as well. Chinese teachers would like to see the one-on-one communication between teachers take place so that they can talk about curriculum design and project implementation. This will significantly enhance the depth and width of teacher collaboration.

Canadian teachers indicated that a significant challenge they faced was the timeline because they felt that it was hard to fit the nature notes projects with other ongoing school projects. They suggested slowing down the pace of the program. They wanted to do one project each semester and focus on one solar term. With these changes, the students would have more time to generate a detailed project ready to be presented and shared with the Chinese side. Canadian teachers suggested that the whole class of students would be able to participate in the Skype meeting or try other possible ways that would allow them to showcase their projects. They believed that more students should have the chance to know about and learn from Chinese students by attending the exchange events. Canadian teachers also suggested that an introductory planning meeting take place prior to each project. In this meeting, both sides can come up with a list of interesting topics so that students from both sides can work on some similar or relevant topics. The connection between project topics will increase the depth of exchange and learning from each other. The last suggestion of Canadian teachers was to have students become pen pals. It will make it much easier for students to learn from each other and increase student engagement with cross-cultural learning.

Implications for Future Practice

The nature notes reciprocal learning program has been a great adventure for both Canadian and Chinese schools for many reasons and benefits. However, it takes time for both sides to figure out a practical way to carry out the goals of the reciprocal learning program. The science team started the program with many different topics of exploration before it narrowed down to the core issue around teacher-led inquiry teaching within the science classroom and student-centered research-based learning outside the science classroom. They did it out of their everyday concerns and interests to achieve a more effective exchange between the two sides, aiming to achieve a mutually beneficial symbiosis between science education and teacher education in Chinese and Canadian schools. The choice of nature notes program gave the collaboration a focus. It was based on the prior experience of successful practice in China and the topic is well situated in today's education movements such as outdoor education (Hofstein & Rosenfeld, 1996), environment education (Zhou, 2015), project-based learning (Tan & Chapman, 2016), and inquiry-based learning (Salinitri et al., 2018). In fact, researchers indicated that the out-of-school project, as an informal mode in science learning, can foster students' learning and help them to become independent and autonomous learners (Hofstein & Rosenfeld, 1996). Robertson (2007) has similar findings. He studied a collaborative lesson for science education, and found that through an out-of-school activity, such as an environmental field trip, the quality of science education is enhanced. Students received cognitive, affective, and social benefits from this learning experience. As the Chinese school stated, the environment is a *dream classroom*, where students can apply what they learned, explore what they do not know, and get opportunities to appreciate nature.

The prior experience of the Chinese school in the nature notes program offered a great resource for the reciprocal learning program to be successful. Curriculum, school support, teacher resource, and community support are necessary factors for the nature notes program to be successful. Canadian teachers felt that there was a time limitation to carry out the program. However, this was not an issue in China since the project was a part of a compulsory course. A lead teacher played a vital role in motivating additional teachers to join the program and serves as a recourse person and mentor to other participating teachers. The support from local government is also important through policies and promotion action (such as the nature notes contests). Having said this, it is necessary to point out that both sides have learned from each other and all appreciate the reciprocal learning experience.

To synthesize the feedback from participating teachers and students, we could like to provide the following suggestions for future similar practices:

Diversifying the Communication Channels In the current project, both sides exchange their work through the limited number of virtual conferences. The time of exchange and the number of participating students were limited. Both teachers and students from Canada and China suggested introducing more communication channels. In addition to the pre-planned exchange events, ongoing communication can be set up between individual students. This can be done through an online platform for students to store and share works, and also conduct chats without a time limit. The technology-based platform enables teachers and students to exchange their ideas effectively and promptly, which means that whenever they have a question, they can discuss and come up with solutions on the platform. Research findings support this point of view as virtual learning communities benefit students' interpersonal collaboration and exchanges (Chia & Pritchard, 2014), and technology can be used to facilitate group communication (Yang et al., 2014). Students can work on the same topics in a small group with partners from their sister school. This will leverage the strength of both sides and generate better intercultural communication and collaboration. Using diverse communication channels will significantly increase the number of students who can benefit from the program and overcome the difficulty of the time difference between Canada and China. The depth of exchange will be enhanced as well since more frequent communication can occur with less limitations.

Coordinating the Projects The current program relies on the graduate research assistants from both Canada and China to coordinate all the exchange arrangements. These assistants have been essential in arranging the meetings and recording student performances. However, it is realized that it will be beneficial for Canadian and Chinese teachers to communicate directly in some ways to share their curriculum design and teaching approaches. As Loveland et al. (2004) indicate, because international collaboration projects are complicated, comprehensive preparation and design are required. Teacher participants all agreed that better collaboration is needed in the future. Direct connection will increase the depth of teachers' collaboration and mutual learning. To achieve this, language is a big hurdle. Graduate assistants are necessary to support language translation. It is also recommended that both sides should work on similar or relevant topics. In the current program, students worked on their own topics of interest around the same solar term, which created student projects that were very different from one another. As researchers have indicated, students effectively obtain information and learn authentic knowledge through the process of doing a project. This process encourages students to take on the role of an initiator of their own learning (Bell, 2010; Tan & Chapman, 2016). If students who have similar interests can initiate work on the same project, they may learn more from each other's works and promote more in-depth questions and learning benefits. The nature note reciprocal learning program was a rewarding experience for participating teachers and students from both countries, although there was space for improvement in the quality of collaboration. Participating teachers and students enjoyed the cross-cultural experiences. They developed a broader knowledge and a deeper understanding of each other's culture. The nature notes reciprocal learning program takes the Chinese traditional culture of the 24 nature notes as an anchor to engage students in rich opportunities of outdoor education. It demonstrates that it is a rewarding experience when students and teachers are engaged in learning in a cross-culture context. Such initiatives should be expanded to other cultural heritages such as the indigenous culture in Canada. The Canadian government has promoted the appreciation of indigenous knowledge (Government of Canada, 2012) and education organizations have worked on the indigenization of their curriculum. Future reciprocal learning around indigenous ways of knowing should be a worthwhile direction to take.

References

Bell, S. (2010). Project-based learning for the 21st century: Skills for the future. *The Clearing House, 83*(2), 39–43. https://doi.org/10.1080/00098650903505415

Chia, H. P., & Pritchard, A. (2014). Using a virtual learning community (VLC) to facilitate a cross-national science research collaboration between secondary school students. *Computers & Education, 79*, 1–15. https://doi.org/10.1016/j.compedu.2014.07.005

Government of Canada. (2012). *Truth and reconciliation commission of Canada: Calls to action*. Retrieved from https://www2.gov.bc.ca/assets/gov/british-columbians-our-governments/indigenous-people/aboriginal-peoples-documents/calls_to_action_english2.pdf

Hofstein, A., & Rosenfeld, S. (1996). Bridging the gap between formal and informal science learning. *Studies in Science Education, 28*, 87–112. https://doi.org/10.1080/03057269608560085

Loveland, T., Miyakawa, H., & Hirayama, Y. (2004). International collaboration in secondary level education. *Journal of Technology Studies, 30*(3), 10–19. https://doi.org/10.21061/jots.v30i3.a.2

Robertson, A. (2007). Development of shared vision: Lessons from a science education community collaborative. *Journal of Research in Science Teaching, 44*(5), 681–705. https://doi.org/10.1002/tea.20162

Salinitri, G., Palazzolo, S., Nahaiciuc, R., Iaeobelli, E., Li, Y., & Zhou, G. (2018). Analysis of Canadian inquiry-based science teaching practices and its implications for reciprocal learning. *Universal Journal of Educational Research, 6*(10), 2280–2293.

Tan, J. C., & Chapman, A. (2016). *Project-based learning for academically-able students*. Sense.

Yang, J., Yu, H., Chen, S.-J., & Huang, R. (2014). Strategies for smooth and effective cross-cultural online collaborative learning. *Journal of Educational Technology & Society, 17*(3), 208–221. Retrieved from https://search-proquest-com.ledproxy2.uwindsor.ca/docview/1556986520/abstract/D951BB43C1E84F47PQ/1?accountid=14789

Zhou, G. (2015). Environmental pedagogical content knowledge: A conceptual framework for teacher knowledge and development. In S. Stratton, R. Hagevik, A. Feldman, & M. Bloom (Eds.), *Educating science teachers for sustainability* (pp. 110–128). SPi-Global.

Index

G. Zhou et al. (eds.), *Science Education and International Cross-Cultural Reciprocal Learning*, Intercultural Reciprocal Learning in Chinese and Western Education,
https://doi.org/10.1007/978-3-031-17157-4

Kolibri 360 GmbH
Ferdinand-Jühlke-Straße 7
99095 Erfurt, Deutschland
produktsicherheit@kolibri360.de